STOP SAVING THE PLANET!

STOP
SAVING
THE
PLANET!

An Environmentalist
Manifesto

Jenny Price

W. W. NORTON & COMPANY
Independent Publishers Since 1923

For information about permission to reproduce selections from this
book, write to Permissions, W. W. Norton & Company, Inc.,
500 Fifth Avenue, New York, NY 10110

For information about special discounts for bulk purchases, please
contact W. W. Norton Special Sales at specialsales@wwnorton.com or
800-233-4830

Manufacturing by LSC Communications, Harrisonburg
Book design by Lovedog Studio
Production manager: Lauren Abbate

Library of Congress Cataloging-in-Publication Data

Names: Price, Jennifer (Jennifer Jaye), author.
Title: Stop saving the planet! : an environmentalist manifesto /
Jenny Price.
Description: First edition. | New York : W. W. Norton & Company,
[2021] | Includes bibliographical references.
Identifiers: LCCN 2020044355 | ISBN 9780393540871 (paperback) |
ISBN 9780393540888 (epub)
Subjects: LCSH: Environmentalism.
Classification: LCC GE195 .P74 2021 | DDC 363.7—dc23
LC record available at https://lccn.loc.gov/2020044355

W. W. Norton & Company, Inc., 500 Fifth Avenue, New York, N.Y. 10110
www.wwnorton.com

W. W. Norton & Company Ltd., 15 Carlisle Street, London W1D 3BS

1 2 3 4 5 6 7 8 9 0

For Danny, Shlomit, Yishai, Jake, Ian, Micha, Gabe, and David, and for your children

The future is yours

CONTENTS

Stop Saving the Planet!

AS A CARD-CARRYING AMERICAN ENVIRONMENTALIST, I'm obsessed with two questions that environmentalists mostly don't seem to be asking—and that I've become convinced are two of the most urgent questions you can possibly ask.

ONE, why aren't we making more progress on environmental crises?

More than half of the 1,340 Superfund sites have been listed for 30–40 years. The air in thousands of communities across the U.S. remains exceptionally dangerous to breathe. And can our climate-change policies even begin to save Miami?

TWO, why do so many Americans hate environmentalists?

How can a green agenda—let's clean up the air and water for your kids!—possibly inspire such widespread rage?

Are you saying, but the problem is big money! It's the ultraright, it's Fox News, it's the oil companies! If so, I don't disagree.

It is also way past time to grapple with the enduring and substantial failures of American environmentalism.

In *Stop Saving the Planet!*, I urge anyone who's having

nightmares about climate change to immediately stop obsessing about how to save "the environment" as a separate world out there.

Stop talking about it, stop hand-wringing about it, stop arm-waving righteously about it—if you want to save people, polar bears, or anything else.

Instead, the most passionate environmentalist question should be:

> How can we *change* environments
> vastly better? And by better, I mean
> a lot more sustainably but also a
> hell of a lot more equitably.

We live *inside* of environments 100%, and we change environments 24/7 to live. Most of us do it ferociously badly and unfairly: that's why the world is 2° warmer[*] than it was in 1900 and why polar bears are starving and why industries pump out over 300 million tons of plastic each year.

Maybe you're objecting, but environmentalists *do* ask that! Not mostly, though, and not really.

Stop Saving the Planet! is about how and why the "save the planet!" approach fails to seriously and sufficiently challenge how most of us live in the 2020s.

I ask why environmentalists find this way of thinking

[*] Fahrenheit, soon to be centigrade

so deeply appealing. I ask why it makes other people so deeply angry.

And I ask what we desperately need to do instead.

I try to be reasonable, if not entirely polite.

Can we please stop saving the $#!ing planet?!

Environment Is in Here

Start protecting our environment today
                                    —earthday.org

IN A NUTSHELL: WE NEED AN "IN HERE," INSTEAD of an "out there," environmentalism.

We need to object vociferously to—instead of double down on—the deeply bizarre assumption that environment is a not-human world "out there."[*]

You change environments to provide food and shelter. You change environments to drive a car, to Facetime with your mom, and to turn on the lights. Your T-shirts, MacBook, Nikes, Ford or Prius, and favorite couch are all 100% made out of wood, oil, rocks, dirt, trees, cotton, animals, and so on.

As humans on earth, we literally change environments to live. Environment is not out there! Your life in 2021 is *foundationally* environmental—whether you live in a 6th-story East Village condo or in the White House or on a farm in Wisconsin.

So yes, to change environments sustainably, we'll need to preserve rivers, wildlife, and lands. Absolutely.

---

[*] however historically powerful that weird Western notion may be

We'll need clean watersheds, and we'll want the bears to stick around. We'll need ecosystems we can all thrive in. We'll have to appreciate, generally, how we *co*-inhabit environments and how everything that's not human responds.

And yet, good luck trying to stop climate change—if you think about it as trying to save, protect, or rescue the not-human environment.[*]

Rather, how can we change environments a thousand times better, to sustain the health and well-being of people and environments?

> We need a Great Changes, not a "save the planet!," environmentalism.

And a Great Changes environmentalism starts with two key premises.

✳ Great Changes environmentalism, key premise #1
We have to change environments dramatically better to create and use our STUFF

Consider how a Prius, for example—as an established "save the planet!" symbol since the 1990s—*changes* environments.

Who cares, in other words, how it saves "the environment" or doesn't emit CO_2?

---

* to wit: Environmental *Protection* Agency

How does a Prius actively contribute to the spewing of greenhouse gases and other pollutants over its 10–20-year life?

Think mining the ore, for example, and growing the cotton, forging the steel, changing the brake fluid, shipping everything, and junking the battery. Your Prius or Tesla, if you own one, facilitates an awful lot of greenhouse gas spewing as you and other people make, ship, market, sell, buy, drive, park, maintain, repair, resell, and finally junk the car.

When you ask how your Prius does pollute—instead of how it doesn't—it can look far less climate-friendly, and a lot closer to your neighbor Tim's Escalade, than its "save the planet!" reputation suggests.

Your Prius, like any car, changes environments wretchedly badly. It contributes a lot more than it doesn't to the wildly high-polluting industrial practices that devastate our atmosphere, bodies, forests, rivers, oceans, wetlands, wildlife, and more.

To create and use our stuff to *not* do that, we're going to have to massively clean up our ultra-toxic industrial practices.

So a Prius? Really, it'll "save the planet"? As an environmentalist strategy, it mostly just "sidesteps the problem."

❋ **Great Changes environmentalism, key premise #2**
We have to change environments dramatically better to create and use our WEALTH

How does your high-mileage hybrid contribute to the spewing of greenhouse gases and other toxics for 10–20 years as it makes money go 'round? And why do you almost never hear a Prius driver ask *this* question?

American environmentalists tend to define environment as a world outside of economy—and even to define and worship "the environment" as an antidote to all our 21st-century profit-mongering.

And yet, what *is* an economy? It's basically how, as a society, we change environments to create stuff and wealth to provide people's needs and wants, and how we distribute the benefits—the stuff and wealth—and the pollution and other costs.[*]

Our economy is *foundationally* environmental. And yes, it's inherently destructive.

When you exchange 31K for a Prius, you fuel a U.S. (and global) economy in which we change environments to maximize profits—most directly, at least, and so often at the cost of the health and well-being of people and environments. The economy, in other words, can be frustratingly hardwired to pollute.

How do you facilitate CO_2 emissions as you earn the 31K? How does the CEO of Toyota spend his share of the profits, and how does he invest it? How much of the money that your Prius generates do the CFO, car dealers, and car wash owners invest in Exxon, BP, and Chevron—among the 20 fossil-fuel companies that are responsible for a third of all greenhouse

---

* even if the definitions in economics textbooks often don't use the word "environment"—which is kinda like trying to define basketball without explaining that it requires a ball

gas emissions—and how do your favorite mechanic's banks, mutual funds, and IRA funds all do the same?

Your hybrid, basically, generates CO_2 and toxics galore as it makes wealth go 'round.

To create and use our wealth to *not* do that, we're going to have to massively redesign the economy to maximize the health and well-being of people and environments.

It's hardly surprising that putting millions of hybrids on the roads for over 20 years has achieved practically nothing at all to roll back climate change or to generally clean up any other unimaginably devastating environmental mess.

And all in all, if a Prius remains an iconic strategy to de-melt the glaciers, then Miami and lower Manhattan are just utterly doomed.

Why, then, do Greenies continue to tout a Prius or a Tesla as a fantastic climate strategy?

Which brings me to the how-convenient if very difficult truth about the "out there" environmentalist approach.

✳ Why "out there" environmentalism?— or, the convenient truth of "save the planet!" thinking

"Life is limitless"

—Prius TV ad

You get to drive a Prius Eco Two or a Tesla Model 3.

You get to feel like you're saving the planet, *and* you can continue to change environments horribly badly.

If you don't really see the scale of the damage—and most of how your hybrid facilitates instead of repairs it—then you don't need to call for massive transformations in how we create stuff and wealth.

You get to have your Prius and drive it too.

Of course, "out there" thinking is especially convenient if you're affluent enough to *have* a whole lot of stuff and wealth. It's especially seductive if you can afford to buy a Prius in the first place and if you live safely far away from the factories and power plants and landfills—and if you can afford to move out of Miami when the seas begin to lap at your door.

American environmentalism has always—always—been haunted by charges of elitism. Environmentalists, in turn, have generally tended to respond with bewilderment: hey, we're saving the planet!

And yet, here, I've become convinced—amidst a blitz of do-little strategies by hybrid enthusiasts, Exxon, and the EPA alike—is the how-convenient truth of "out there" environmentalism:

> You can genuinely care about our urgent environmental crises—but also fail to seriously challenge the über-toxic industrial and economic practices that create and perpetuate them.

On top of all that, you get to feel like you're a downright eco-hero—or that you're doing *something*, at least.

Which brings me, finally, to Green Virtue and Whole Planetude—the two powerful "save the planet!" credos.

✳ How "out there" environmentalism works to legitimize itself

Save paper, Save the planet
> —sign above paper towel dispenser,
> Princeton University
> environmental studies building

Green Virtue and Whole Planetude are environmentalist instincts, almost. It'd be hard to exaggerate their influence. And they both flow logically from thinking about environment as a not-human world "out there."

Green Virtue—I recycle, I'm an amazing person!—is the greener-than-thou-ness that has always plagued environmentalism. Alas, it's rooted in a long Western tendency to exalt the environment "out there" as the *real* world—pure, true, and unsullied by human corruption.

What could possibly be more virtuous, after all, than rescuing the real, uncorrupted world?

Green Virtue can be spellbinding—and half-persuade you to imagine that your Tesla or Prius soars heroically into the sky at night to scrub carbon out of the atmosphere.

In the first half of this book, I track how, as "save the planet!" credo #1, Green Virtue powerfully legitimizes a wide range of popular "sidestep the problem!" (or even "screw the problem!") strategies—from runaway green consumerism to

LEED-certified corporate HQs to generous public subsidies for both.

And Whole Planetude, which I turn to in the second half? It assures you that all environmentalist actions achieve the same ultimate goal—to "save the planet!"—and therefore that all actions are effective.

Buy a Prius, eat an organic peach, regulate CO_2 emissions. It's all good. Whole Planetude assures you that your hybrid does *something*—so whether your Prius is significantly different from an Escalade doesn't actually matter.

Whole Planetude, too, infuses and exalts a lot of popular "save the planet!" strategies—hybrid buying, greenwashing, pollution offsets and trading—that sidestep the problem and that have failed for decades to save much of anything at all.

✳️ Am I saying that all American environmentalism takes the "out there" approach—which allows you to sidestep the problem, and to have your Prius and drive it too?

NO!!

No, I am not.

The "in here," Great Changes approach to environmental crises is wonderfully alive and well and kicking very hard—if a lot less likely to appear on a T-shirt.

A great many environmental advocates—individuals, communities, NGOs, scholars, companies, and public agencies—are figuring out how we can change environments to create and use our stuff and wealth dramatically more sustainably and equitably.

They're not necessarily shouting that we have to "save the planet!"

They might just save you, me, your neighbor Tim, your grandkids, the polar bears, the grandbears, and Miami alike.

And I showcase some of their work in my 39 Ways to Stop Saving the Planet.

In the meantime, it can be useful to ask—Why do so many people hate environmentalists?

2nd Reason

Everyone Hates Environmentalists

Smug Alert

—*South Park* episode, 2006

OK, NOT *EVERYONE*.

And yet, why does environmentalism provoke so much annoyance, alienation, and downright rage?

Green Virtue, of course, explains some of it. The notion that you're Captain Planet—or a mighty Save the Planeteer, at least—because you're saving "the environment" (from all humanity, no less) is just notoriously annoying.*

And yet, how often do you hear these three gripes as well? Environmentalists don't care about the economy! They don't care about people! They're elitist: they don't care about the *little* people!

The "out there" environmentalist approach, I think, aggressively fuels all four of these common complaints.

And it does so exactly because it refuses to insist on

---

* even to a lot of environmentalists

the central importance of environment to people's lives "in here."

✳ Yes, environmentalists care about the economy

> *Our . . . environmental . . . policies balance America's need for a healthy . . . environment with economic growth*
>> —Obama administration's energy.gov site

How frustrating, though, when they insist that we have to "balance" the needs of the economy and "the environment"—which betrays an "out there" understanding of economy and environment as two separate and often opposing forces.

This "balance" mantra just throws gasoline on the anti-environmentalist beef—by CEOs, miners, and Amazon warehouse workers alike—that environmental regulations hamper economic growth.

Also, it's utter bullshit. So how about insisting instead that, hey, if you're struggling in this economy, then it's decidedly *not* because the air your kids breathe is too clean or because the water they drink isn't poisonous enough.

No, it's because the goal of our economy is to maximize growth and profits—*not* your health and well-being—and we achieve this goal by changing environments hugely badly and unfairly.

Basically, the way our economy works is that we use high-polluting industries to make and deliver goods and energy as cheaply as possible. The wealthiest Americans reap the lion's share of the profits—and we stash the lion's share of the devastating pollution in the communities where people are struggling most of all.

How about saying that our economy (like any economy) is foundationally environmental—and that a healthy economy for all requires healthy environments.

If you're a miner or if you work in an Amazon warehouse, then you earn decent to poverty-level wages in unsafe working conditions to help blow up mountains and spew toxic chemicals—which we stash in your communities.

If you're the oil-industry titan Charles Koch?—or Jeff Bezos, the CEO of Amazon and the world's richest person? Then yes, our Great Changes environmentalism—which calls for a wholesale transition to an economy that is fair and sustainable*—will threaten your outsize, gargantuan profits.

We need a lot more, not less, environmental protection.

How about saying that instead?

✳ Do environmentalists care about people?

New Prius helps environment by killing its owner
                                    —*The Onion*, 2012

We do! And yet, environmentalists often shout—unhelpfully—that we have to save "the environment" "out there" from *what*? People! All of us, as a species!

---

* Green-New-Deal-style. See reason #11.

Humans suck, is so often the message.

How about saying instead that if we change environments "in here" dramatically better, we can vastly improve millions of people's lives while also protecting the environments we love and rely on.

❊ Are environmentalists elitist: do they care about the little people?

Social reform will not protect the climate
—*Washington Post* editorial
on the Green New Deal, 2019

Well, environmentalists have notoriously tended to ignore the vast inequities in how we change environments—not least because they obsess most of all about how to save the environment "out there."

Still, why isn't this gripe red-flag troubling to any environment lover who wants to clean it all up—especially since these "elitist!" accusers include the very people who suffer the most from environmental messes?

I mean, it's easy to understand why the gazillionaire Koch brothers and other ultraright wealthy American elites hate environmentalists—as in this economy, *affluent folks reap their riches precisely by changing environments wretchedly badly.*

It's easy to explain why the Trump administration attacked environmental regulations more vehemently, arguably, than any other target*—and why "energy & environment" topped the über-right Heartland Institute's "Action Plan for President Trump."

And yet, the haters also include two large lower-income demographics—who, tellingly, tend not to agree politically on much of anything else.

They include a great many lower-income Americans of color—who on average suffer the most devastating pollution of all. They also include a lot of the white lower-income right-wing Republican base.

Why are these least affluent Americans, who are most likely to live by the factories, landfills, and Superfund sites—and who suffer crippling rates of asthma, cancers, and other environmental illnesses—so convinced that environmentalism is not about *them*?

How about insisting that this is one of the most urgent questions that anyone who is genuinely freaking out about environmental crises can possibly ask.

✳ 1, 2, 3, ka-boom!

Earth suffers as climate change is ignored
—*New York Times*, 2015

---

* 1st day post-election: appointed EPA foe to head EPA transition, vowed to bail on Paris climate agreements. 1st day post-inauguration: scrubbed climate-change info from official websites. 1st-budget target for biggest cuts: the EPA.

What good is it to save the planet if humanity suffers?

—Exxon CEO (& now ex-secretary of state) Rex Tillerson, 2013

In the meantime, the "out there" approach creates a huge opportunity for the Kochs and ultraright politicians alike, who stir these complaints together and proceed to dip meretriciously into this stew in exceptionally damaging ways.

Corporations that see regulations as a threat to their outsize profits declaim: These greener-than-thou tree huggers don't care about you or your jobs! Right-wing pundits shout: Climate change is an elitist plot!

In fact, what we're seeing, after five decades of saving "the environment" . . . well, what we're seeing is the culmination of a long history of anti-environmentalist antagonism. And it's now virulent, mobilizing, and extraordinarily hateful.

"Environment" has become *the* dirty word in 21st-century American politics—which makes climate change the f-word—and is arguably the #1 right-wing code word for "liberals/Democrats don't care about *you*."

And "out there" environmentalists make these attacks so easy—with every "save the planet!" T-shirt and grocery bag, and with every explanation by a senator or EPA official or NGO that reducing vehicle emissions will "help the planet" or "save the environment."

With every call to "balance the needs of the economy and the environment." And with every *New York Times* headline that reads, "Earth suffers as climate change is ignored"—

which allows Exxon's CEO to swoop in and ask, what good is it to save the planet if humanity suffers?

Climate scientists and many others have often called on journalists to explain the beyond-911 urgency of the climate crisis.

Yet what we have here isn't a failure to explain or to communicate.

No, what we have at this point is an all-out class and cultural war that's now coming to a head, but that's been raging around American environmentalism from the very beginning . . .

> . . . in which a hypervirtuous crusade to "save the planet!" has failed to insist on the centrality of environment to the economy and our everyday lives . . .

. . . and has also been seriously getting on people's nerves.

How about saying, Yeah, it's time to have a different conversation about environment.

I Can't Solve the Middle East Crisis by Myself

50 Simple Things You Can Do to Save the Earth

101 Ways You Can Help Save the Planet before You're 12!

Eco-Horsekeeping: Over 100 Budget-Friendly Ways You and Your Horse Can Save the Planet

— a few of the 95+ "save the planet" books
on amazon.com

WHY CAN'T YOU FIND A *50 SIMPLE THINGS YOU Can Do to End World Poverty* handbook at your bookstore?—or *101 Ways You Can Help Stop Gun Violence (or Solve the Middle East Crisis) before You're 12!*

Or maybe *100 Budget-Friendly Things You and Your Horse Can Do to End Slave Labor Worldwide.*

Why track your everyday actions through climate change and the plastics Armageddon—but not through any other global monumental fubar crises? Is global warming easier for you to fix than world poverty? Maybe, maybe not. Are you more personally responsible for the Great Pacific Garbage Patch than for what's happening in the Middle East? It's debatable.

Hello again, Green Virtue! Which flows logically from "out there" environmentalist thinking and also deeply infuses the most popular strategy to "save the planet!"

To do it yourself.

Am I saying, keep all the lights on in your house? Toss all your blue bins into the black bin? And by all means, purchase an old Hummer and remove the emissions controls and drive it back and forth from Brooklyn to Portland?

Absolutely not!

I am saying, an "in here" environmentalism insists on a half-dozen urgent Howevers.

And I'll start with Green Virtue itself.

✳ However 1
Can we please, please already finally dial down the greener-than-thou mishegas?*

Do you say "I'm awesome!" when you decide not to park in a disabled space? Do you feel oh so virtuous when you choose not to cut in front of other people in a supermarket line?

How about when you pay taxes every year?—which we call just having a social contract.

Call this collective virtue—which most of us take for granted, and which we, as a society, often legally enforce.

* [Yiddish] blarney, malarkey, poppycock, horsefeathers, hogwash

Why can't environmentalists just be good citizens instead of righteous Save the Planeteers?

Also, why is recycling a personal choice to begin with?

Which brings me to However 2.

✳ However 2
The passionate emphasis on individual actions can blind environmental advocates to the importance of larger public actions

Take waste, for example. In the U.S., all municipal (versus industrial) solid waste adds up to about 3% of the total waste stream. All that sorting, composting, and cloth-bagging. All that squinting and wishing and hoping while you're trying to read the # on the bottom of the plastic thingamabob. 3%.

Or consider hybrid cars, which accounted for about 2% of vehicles on U.S. roads in 2019 and actually peaked with gas prices at 3% in 2013. Again, 3%. And just on U.S. roads.

"If everyone would recycle, or compost, or fly less" is like a Greenie mantra. And it's like sending a carrier pigeon to the firehouse instead of calling 911.

> The "if everyone would do it" mantra tends to wildly overemphasize the impact of your personal choices—and to dangerously underemphasize the dire urgency of public and systemic solutions.

What we really, desperately need is "definitely everyone," "collective virtue" policies, laws, and regulations—to require *all* cars to be low-emission, *all* paints and carpets to be non-toxic, and *all* waste to be reduced (mostly), reused, and/or recyclable.

❋ However 3
The emphasis on the importance of individual actions conveniently passes the buck from producers to consumers

Are you really trying to stop the Noachian flood of plastics, for example, from the consumer end? You're not even sending a carrier pigeon—you're dispatching a garden snail—instead of calling 911.

Why shouldn't Ford, Toyota, and BMW prioritize low emissions and high mileage? Why shouldn't Apple and Samsung use less toxic materials to manufacture their phones, and try to make the phones last as long as possible, and take responsibility for where all the toxics in the millions of obsolete phones end up?

Why does Whole Foods urge their *customers* not to use plastic bags?

Well, at the moment, the American economy is hardwired to change environments to maximize wealth more directly than anything else—often in the almost quasi-religious belief that an economy that maximizes growth and individual profits will somehow magically maximize health, well-being, opportunity, liberty, democracy, and so on.

Companies aren't expected primarily to offer goods and services to sustain the health and well-being of people and environments. They're legally encouraged and in some cases required* to maximize shareholder profits as the primary goal.

Consumers can demand products and services that sustain health and well-being—and thereby can make more ethical business practices profitable.

And **government** plays a truly bizarre Janus-faced role. It encourages and facilitates companies to maximize profits—while it also protects the public (think pesticide bans or emissions standards) from the inevitable and disastrous consequences.

In other words, the 21st-century U.S. (and global) economy largely outsources virtue to individual consumers and NGOs—and to governments, which often just design subsidies, tax breaks, and other incentives to pass the buck to virtuous consumers.

* potato, pot-ah-to, it's heartily debated—but the company has to maximize its interests, which corporations since the 1980s have increasingly defined as shareholder wealth

And what consumers could possibly be more eager to consume virtuously than Save the Planeteers?

Consider, as exhibit A, the much-beloved Keep America Beautiful anti-littering campaign—which Coke and other companies launched in the 1950s. It's failed spectacularly to accomplish anything except to move the ever-growing trash crisis from the sidewalk, where you can see it, to the landfill, where you can't.

Likewise, when Congress moved to ban and limit nonreturnable packaging in the 1970s, Coke led the fight to establish municipal recycling programs instead—and thereby successfully continued to rebrand the "producing toxic crap" problem as your "consumer waste" problem.

And by 2021? About 8–9% of plastics actually get recycled. Your average fish ingests plastic fibers by the hundreds of thousands annually, before you and your kids and your (and the Coke CEO's) grandchildren ingest the fish.

And our recycling programs have collapsed spectacularly under their "pass the buck" weight—as China, India, Cambodia, and other countries increasingly refuse to take Americans' plastic and other garbage.

Which has inspired Coke[*] to launch a huge new PR campaign to encourage you to recycle.

Why not call 911 instead?

Why shouldn't we, as a society, expect and require Toyota and Apple and Coke to create good, durable, low-toxic

---

* which, with other companies, has *always* known how difficult it is to recycle plastics

stuff—and to generate profits first and foremost by changing environments cleanly, sustainably, and equitably?[*]

※ **However 4**
When you single out eco-crises as the only crises for which you're personally responsible, you might ignore how they actually fuel—and are fueled by—urgent social, economic, and political crises

You might not see how the 2010 Deepwater Horizon oil spill continues to devastate coastal economies in Louisiana, Mississippi, Florida, and Alabama.

You might fail to ask how the destruction of wildlife habitats makes human pandemics more likely, or how devastating air pollution has made low-income communities far more vulnerable to the COVID-19 pandemic.

You might not see how local opposition to energy extraction by multinational companies fuels protests and violence across Latin America. Or how the inequitable use of oil resources fuels ISIS. Or how ivory poaching and illegal logging finance child-soldier armies in parts of Africa.

You might fail to ask, generally, how the mismanagement of natural resources fuels wars and conflicts globally. How the lack of access to clean air and water—and to healthy food and to parks and green space—fuels poverty, obesity, and violence.

* wouldn't that be, like, more efficient?

And how to be poor in the 21st century—in the U.S. and often far more dramatically outside it—is to be sick.

> All environmental crises are social, economic, and political crises—and vice versa.

You might fail to think about how most crises world-wide have been and continue to be impossibly tied up with how economies, governments, corporations, and individuals change environments badly and inequitably.

✳ **However last but not least**
The "if everyone would do it" approach tends to ignore and flatten the vast disparities in who actually creates and perpetuates the crises

Swedish-pancake flatten.

It tends to ignore and minimize an abundance of dire and essential "who" questions. Who creates the messes? Who benefits most? Who suffers the worst consequences? Who should bear more responsibility for cleaning it all up?

For example, let's say Sam doesn't recycle. His job is to flip burgers at Burger King, and he and his family share a 2-room apartment with his cousin's family. Is Sam really destroying the planet?

Tom, however, is a lawyer who sits on Burger King's board of directors, and he virtuously composts all his leftover Whole Foods veggies in the backyard of his 17-room house with 5 energy-efficient bathrooms. Is Tom really saving the environment?

"What can *I* do?"

I have been urging people—family, friends, students, colleagues, audiences, and random people at parties—to please, seriously, just please stop asking this question first and foremost.

And what do you think the first reaction is? "OK, I get it, but what *can* I do?"

Almost always.

> "What can I do?" is an important question—but only when it doesn't prevent you from asking, "What needs to be done?"

Yes, of course, there are things you can do and should do—39 life hacks, to begin with, just in this book.

And yet, it's essential to recognize that the intense obsession with "what can *I* do" has, for a very long time, been a very large part of the problem.

4th Reason

Are You Really Trying to Save Energy by Using More Energy More Efficiently?

Come see my _____ new house
with natural _____ and sustainable _____
and efficient _____! We'll eat _____,
GMO-free _____, and organic _____
with locally sourced _____ in the solar
_____ in the _____ wing built with
_____, nontoxic _____, and _____
materials, and my all-time favorite feature—
a _____ _____.

✳ **? Riddle Me This ⸞**

**Can you save more energy by buying 1
Prius and driving it 15 miles or by buying
3 Priuses and driving them 5 miles each?**

I'll come back to that . . . but you can now buy organic
water. You can buy vegan non-GMO organic gummy
bears, organic ear-stretching cream, earth-friendly bul-

lets, carbon-neutral condoms, and reclaimed-wood wedding cake stands.

You can buy a Samsung Energy-Star-certified 82-inch TV, and you can cough up $1,900 to spend a night at 1 Hotel Central Park under a reclaimed barnwood ceiling. In 2012, you could buy $900 Manolo Blahnik high-heeled sandals made from discarded tilapia skins. Right now, you can hire an eco-concierge to help you shop green, book an eco-trip, and buy carbon offsets.

You might like to buy a $162,000 hybrid Porsche SUV that gets 17 mpg—or a $100 million eco-friendly hybrid 274-foot Sparkman & Stephens mega-yacht.[*]

Hi again, Green Virtue!

In the 21st century, we've seen an explosion of green consumerism: a market-shaking, mall-quaking, Internet-swarming, PR-breeding supernova explosion.

And if "what can *I* do?" has been the most popular Save the Planeteer question, the most popular answer arguably has been, "buy something!!"

At best, hooray. Any consumer product should of course be as nontoxic and energy-efficient as possible. Why *not* sustainable wedding cake stands and hybrid Porsche SUVs and carbon-neutral condoms?

At worst, this "buy something!" craze tries to use the problem to solve the problem.

---

[*] or you can charter the yacht for $1,120,000/week

> It assumes, conveniently, that we can use runaway consumerism to clean up the messes that runaway consumerism creates.

And it fails by a moonshot to challenge the profit-maximizing economy, which requires the vast and growing production and consumption of STUFF.

❋ ? Riddle Me This ⸴

Why do so many Americans seem to believe that we can shop our way out of climate change (other than, wow, how easy and fun would that be)?

Well, here's the thing about our mega-consumerist economy: it fuels an über-enthusiastic consumer *culture*, in which buying stuff means infinitely more than just paying for stuff and using it.

We buy mountains of stuff to define and entertain ourselves and to nourish our social relationships. You might go shopping with friends to have fun—call it Saturday—or you might hop online to shop when you're bored. Me, I feel an instant kinship with folks who wear Chaco sandals, and my mom bought me my favorite purse at an art fair.

Here's another thing about mega-consumerism: a lot of consumers think consumerism sucks. At least, you or I might not be wholly enthusiastic. Are you suspicious of this buy,

buy, buy culture? Are you tempted by Buy Nothing Day as the Black Friday alternative? Can it all feel wasteful, superficial, and meaningless?

Welcome to a long American tradition—of ambivalence about mega-consumerism. Which fuels an explosive subculture of virtuous mega-consumerism—an equally established tradition.

You buy stuff to solve the world's problems—the more stuff, the more good you'll do!—and you assuage your anti-consumerist suspicions while doing it. Buy a doll, and the toy company donates another to a children's hospital. Buy a half-gallon Mega Jug soda, and your local KFC donates $1 to the Juvenile Diabetes Research Foundation.[*] Buy fair trade, union-made, cruelty-free, made in America.

And again, who could possibly be more eager to consume virtuously than Save the Planeteers? Buy organic, all-natural, recycled, LEED-certified, vegan-certified, GMO-free, Energy Star, sustainable.

How much of green consumerism is about being green—and how much of it is about feeling virtuous?

Or as the strangely honest Chevy Volt ad confesses: "Electric when you *want* it, gas when you need it."

✳ ? Riddle Me This ⸘

So how, exactly, does Green Virtue consumerism play out in our mega-consumerist economy?

[*] not making this up, it happened

What consumers do: You and I and Gwyneth Paltrow consume untold tons of energy and resources in order to save energy and resources. Buy 3 hybrids, and you'll save 3 times as much energy!

Also, we toss out mountains of older stuff that works just fine—the 2-year-old not-a-hybrid car, the "destroying the planet" light bulbs, the "makes me feel guilty" not-Energy-Star refrigerator.

EXAMPLE: You trade in your 2-year-old Toyota Corolla for the latest Prius hatchback or Tesla Model 3—but now you have to cancel out the carbon footprint (actually higher than for non-green cars) of manufacturing the brand-new car.

Just the carbon footprint, and just the manufacturing—and now, add all the energy and resources the car gobbles up throughout its lifetime, and all the CO_2 emissions and toxics it generates—as a physical object, but also as an economic object that makes money go 'round.

What a lot of companies do: They gobble up energy and resources to create, market, and sell as much green stuff as they possibly can.

Also, they greenwash, to persuade you of their Green Virtue.[*]

What government does: We create policies, programs, and tax incentives to encourage you and me and Ben Affleck to virtuously buy tons of new green stuff and to throw away tons of perfectly good less green stuff.

* see reason #5

EXAMPLE: Cash for Clunkers programs*—such as the version that the Obama administration pitched as a two-fer on the heels of the 2008 crash, to stimulate the economy and to "save the planet!"

Here's how it worked. You got a subsidy of up to $4,500 if you traded your 18-mpg-max vehicle for a new vehicle that got at least 22 mpg. Or you could trade in a 16-mpg-max SUV for a vehicle that got at least 18 mpg.

Or you could trade a 14-mpg-max SUV or pick-up for a vehicle that got at least 15 mpg. Read that again . . . and this proved to be the most popular swap.

Also, your "clunker" had to be less than 25 years old, and you had to buy a brand-new car. And the dealers were required to destroy the "clunkers"—so the program was actually shredding 3-year-old pick-up trucks.

How could this program possibly not increase CO_2 emissions? On the contrary, the environment-friendly Obama administration deployed Green Virtue primarily to boost our infinite-growth economy—which is about as likely to slow climate change as to make hell freeze over.[†]

✳ ? Riddle Me This ¿
Huh?

* & carpool-lane permits for hybrids, & tax breaks for new LEED-certified construction, & so on

† which itself is now even less likely to happen, what with climate change & all

Buy stuff. In moderation.

There's too much stuff, though.

Runaway green consumerism encourages Toyota and the EPA and you and me and Ben and Gwyneth to justify the often mind-blowingly unnecessary consumption of energy and resources that will always be, and is by definition, thoroughly and heart-stoppingly unsustainable.

And to feel good for doing it.

Greenwashing Works

Take your recycling to the curb in a bag that's
strong enough to handle it
                    —Glad Trash Bags website

12 Koch companies sites are @WildlifeHC
certified for #environmental stewardship
        —Earth Day tweet, Koch Industries, 2015

GREENWASHING, ACCORDING TO THE *FINANCIAL Times Lexicon*, is "the overstating of the environmentally or socially conscious attributes of a firm's offering and the understating of the negative attributes for the firm's benefit."

Why does it work?

I mean, how many of us really see Nestlé's ECO-Shape® bottle with 22% less plastic—which the company pitches as a heroic response to the disastrous proliferation of plastic water bottles—as a truly genuine or effective solution?

Apple's LEED* Platinum HQ in Cupertino and its

* Leadership in Energy & Environmental Design—the certification program that the nonprofit U.S. Green Building Council has run since 1998

four U.S. data centers use 100% renewable energy. The company also stamps "printed on recycled paper" on every tiny gift card—in step with the uniquely American compulsion to make sure you know.

Hi again, Green Virtue—annoying even to environmentalists, and yet still often so irresistible.

Companies deploy it effectively—it's like holding up a "we're so virtuous!" sign—to proclaim that they're "saving the planet!" but also that they're truly wonderful all around.

Nestlé aggressively drains and privatizes public water sources to fill its zillions of ECO-Shape® bottles* in the U.S. and also in countries that don't have sufficient drinking water. The company refuses to weed out contractors who use slave labor and lobbies fiercely against the proliferating efforts to make access to water a universal right.

Apple, to be fair, boasts aggressive recycling and clean energy goals. And yet, the company builds and operates its decidedly non-LEED factories in China, where they can skirt U.S. labor laws and where they're free to pollute at drastically higher levels. Apple works hard to make its devices obsolete. And a great many MacBooks and components still end up in huge carcinogenic heaps in Ghana, China, India, and many other countries.

* allegedly illegally in some places

In the "sidestep/screw the problem!" lexicon, I'd tweak the *Financial Times* definition.

> Greenwashing is the deployment of small acts of Green Virtue—which allows a company to continue to change environments mostly badly and inequitably to maximize growth and profits.

Consider Walmart's vaunted eco-heroics since 2005—when the CEO and board of directors set out to redeem the company's increasingly lousy reputation, along with its falling stock prices.

What did they decide to do? They announced a big sustainability initiative.

What *didn't* they do? They didn't substantially redress the notoriously low wages and benefits, the egregious pay and working conditions, or the bullying of suppliers. They didn't forsake any of the dozens of offshore tax havens where the company stashes tens of billions of dollars. Or stop contracting with abundant companies that use prison labor. Or stop making large contributions to climate-denier PACs and politicians. Or stop bribing foreign governments. Or stop spending millions of dollars to lobby against labor and environmental regulations.

They didn't increase the average quality or durability of their other products.

They sold LED light bulbs and organic vegetables, installed rooftop solar panels, and retrofitted their vehicle fleet to be more fuel-efficient.

Almost all of which has saved them a lot of money—and has bumped up the company's appeal to affluent green consumers.

Walmart has since increased their carbon emissions, and also remains one of the world's biggest polluters.

And you can readily interpret their sustainability initiative as a very smart and strategic—and very "sidestep/screw the problem!"—deployment of Green Virtue.

Are you saying, Hmm, yeah, but? So OK, here are two popular and perfectly legitimate HʏBs:

❋ **Hmm yeah but #1**
   Maybe Apple and Walmart and Nestlé and Exxon* are genuinely trying to green their companies?

Maybe, maybe not. Or maybe both. A "save the planet"-minded CEO—or a company's sustainability officer—might be deploying Green Virtue furiously and strategically while also conscientiously identifying small green things to do.

* green company of the year in 2009, according to a *Forbes* cover

❄ Hmm yeah but #2
Isn't it great that Walmart, Exxon, and other global behemoths are doing something, at least— instead of nothing at all?

Maybe, but not necessarily. It depends. You really have to tally the gains and losses.

You can look at Walmart's green efforts and say bravissimo. Yet building a LEED data center or installing rooftop solar panels is about as likely to stop the glaciers from melting as hanging a huge "we're so virtuous!" banner across the entrance to the company headquarters.

And if Walmart fails to do anything it actually needs to do to make a measurable difference, then who cares, honestly, about its organic spinach?

Tally it up, tally it up. How do a company's climate initiatives work to legitimize business as usual and therefore to actually perpetuate climate change?

In the last decade, the LEED Platinum corporate headquarters has emerged as the new corporate green must-have. Exxon, BP, Chevron, Nestlé, Apple, McDonalds, Bank of America, Goldman Sachs, JP Morgan Chase, Georgia-Pacific, PepsiCo, Clorox, Weyerhaeuser, and the chemicals behemoth BASF all now have one.

Bank of America touts its own LEED HQ, the Times Square–adjacent One Bryant Tower, as especially state-of-the-art.[*] Yet in 2019, Bank of America financed fossil-fuel-

---

[*] & in 2013 at least, it was using more energy and emitting more greenhouse gases than any similar-size NYC office building

industry companies to the tune of $48 billion—which those companies then spent in part to fight ferociously against climate and pollution regulations.

Does Bank of America really deserve applause for its LEED Platinum HQ—where the CEO, the CFO, and a handful of the bank's employees breathe the cleanest air in midtown Manhattan?

Reason

Everyone Hates Environmentalists (Add It Up)

Where we live, work, and play
>—environmental justice definition of environment, coined at the 1991 First National People of Color Environmental Leadership Conference

THE OBSESSION WITH "WHAT CAN *I* DO?"

The popular answer, "buy something!"

All the corporate green-itude.

And the Green Virtue that powers so much of all of it.

What does all this "saving the planet!"[*]—as the environment "out there"—look like from a coal town in Appalachia, or from Southeast L.A., or the industrial South Valley in Albuquerque, or impoverished rural counties in Alabama, or the Yakama Reservation in Washington, or working-class neighborhoods in St. Louis and Atlanta and Cleveland and Chicago and Boston?

[*] see reasons #3-5

✳ Start with the problem

Let's say you live in Southeast L.A.—which maybe you do. If you don't, you might not know that it's a 135-square-mile area in south L.A. County. It's predominantly Latino and lower income. And it suffers the worst air pollution in the L.A. area—which itself has long boasted the worst air pollution of any U.S. metropolitan region.

The most heavily industrialized urban area in the country, Southeast L.A. straddles the 710 freeway, which connects the combined L.A./Long Beach ports—the largest U.S. port complex, and the single largest source of air pollution in Southern California—to the rail yards and warehouses in downtown L.A.

Cargo trucks spew diesel fumes 24/7, as do the Brobdingnagian container ships docked at the port. The dense smaller-scale manufacturing produces an abundance of toxic emissions and dump sites, and the auto, aerospace, and other post–World War II mega-industries left a raft of Superfund sites behind when they pulled up stakes in the 1980s and '90s.

Asthma, heart disease, cancer, premature births: the Angelenos in these residential neighborhoods suffer some of the highest rates in the country—and in August 2020, this was the COVID-19 epicenter in L.A. County.

In other words, if you live and/or work in Southeast L.A.—in and next to the factories and freeways and waste sites—you bear the brunt, dramatically and disproportionately, of how we, as a society, change environments wildly badly to sustain the goods-intensive, consumer-oriented, growth-dependent, profit-maximizing global economy.

And you would think . . . well, if we were really, seriously making substantial progress in the U.S. on cleaning up environmental messes? You would think we would especially see the results *here*.

So do you?

It sure doesn't feel like it.

✳ Add the popular individual and corporate solutions

And yet, environmentalists tell you incessantly that all humans are destroying "the environment" everywhere.

And not just that, but apparently you can "save the planet" if you buy tons of expensive green stuff that you mostly can't afford.

And not just that, but all these hybrid SUVs and toxic-free carpets and organic GMO-free sour gummy bears are cleaning up other people's (and other people's kids') homes and bodies, in affluent neighborhoods that are generally far less polluted than yours.

And not just that, but all this SUV driving and gummy bear eating is accompanied by a never-ending greener-than-thou drumbeat, about how supremely virtuous environmentalists and their kids are because they buy and use all this stuff you can't afford.

And not just that, but the energy, waste, and other high-polluting companies—who can't seem to figure out how not to dump devastating toxics into your communities—likewise kvell incessantly about how ultra-virtuous *they* are because of their shiny new state-of-the-art LEED Platinum HQs, with high-tech air filters and bike parking and low-flow toilets and rooftop solar panels.

And not just that, but you and your kids, who all suffer from asthma, are just *not* so virtuous, apparently—you're part of the problem, not the solution—because you're not buying tons of green stuff you can't afford to save "the environment" from humans everywhere.

You!

You, amid the diesel fumes and smokestacks and Brobdingnagian port cranes and freeways and Superfund sites and landfills and the just everyday toxic dumps that make this global economy—which by all means includes the supernova hosanna-shouting green-consumer economy—*possible.*

> Here, in Southeast L.A., a lot of the "save the planet!" frenzy looks like silly elitist ridiculous nonsense.

It looks self-indulgent, blind, hypocritical, often callous, and just spectacularly clueless.

A lot of it is just outright farcical. You know, it'd be funny, except it's not.

(And if you're watching and experiencing all this from the Niger Delta, where BP, Shell, and/or Exxon has virtually destroyed the air, water, and land in the process of extracting oil and gas and has made your communities wholly unlivable? It just seems Kafka-level batshit.)

❋ Add "save the planet!" public policies

Meanwhile, your federal and local governments offer tax breaks to affluent consumers to buy the hybrid SUVs and other expensive green stuff.

And not just that, but many "out there" environmentalist policy wonks—whether at universities, NGOs, or public agencies—outright *disavow* the importance of prioritizing cleanups in lower-income communities (like yours) in the U.S. and abroad.

NGOs and public agencies, to begin with, often insist that social justice and environmental devastation are entirely different crises. They say, our job is to focus on climate change.

All while climate policy makers, for their part, tend to call for CO_2 reductions anywhere—and therefore tend to entirely ignore the devastation that the co-pollutants (sulfur, mercury, particulates, and so on) wreak on the residents in your communities.

❋ The upshot

If you live in Southeast L.A., you contribute least to creating the messes. You benefit least from how we change environments to create stuff and wealth. You suffer the worst of the ultra-toxic consequences where you live, work, and play. You generally benefit least from the solutions. And you're castigated for not being virtuous enough.

And environmentalists wonder why you might find environmentalism infuriating.

Since the 1970s, environmentalists have liked to show "before and after" pictures—to remind us of how much worse "we" were before the movement swept the country in the 1960s and '70s. And true, the air and water are much cleaner in many places.

Typically, however, it remains enormously unsafe to breathe the air or to drink from the kitchen tap in low-income areas across the country.

> Since the 1960s, the U.S. environmental movement has too often played out as a trickle-down initiative—which cleans up environmental messes most successfully where affluent Americans encounter them.

American environmentalism has *always* been haunted by a Great Class Divide—which right now in 2020 adds up to three huge barriers to tackling environmental crises.

One, "save the planet!" actions and policies—by individuals, companies, and the public sector—are failing spectacularly to clean up the places with the worst messes, where people suffer the most devastating consequences of air and water pollution, climate change, and other crises.

Two, this consistent failure to act in the interest of lower-income communities—predominantly communities of color and also abundant white communities—has profoundly sabotaged the broad public support we need to clean up the devastating environmental messes.

And finally, if environmental actions and policies are failing dramatically to clean up the places with the worst messes, then how much progress are we really actually making?

Are we making progress?

You can't answer that question while standing next to your Prius hatchback in the driveway of your solar-powered house in Santa Monica.

Drive the Prius down the 710 freeway.

Come to Southeast L.A.

Are we getting enough done?

Can You *See* Any People on the Blue Marble? (Just Add Whole Planetude)

People + Planet
—Aspiration financial firm newsletter

HAVE YOU SEEN THE ● RECENTLY?— AKA the "whole Earth," aka the Blue Marble.

Or maybe you've moved to Mars.

The "earth from space" image is astonishing, for sure—and yet, as the beloved "save the planet!" icon,[*] the "whole Earth" reinforces the conviction that, ultimately, the goal of any and all environmental actions is to "save the planet."

"People and the planet": have you heard *that* recently?[†] The common use of "the planet" as an

---

[*] as of 9/20, the pic for the Wikipedia pages Environmentalism, Environmental Movement, Sustainability, Ecology, Deep Ecology, The Climate Reality Project, Environmental Studies, Environmental Science, Environmental Humanities, the Ecology portal

[†] or maybe you've moved to Neptune

exact synonym for "the environment"—to mean everything on the planet but humans—just doubles down on the "out there" vision of environment as a unitary, not-human world.

In the first half of the book, I tracked how the "out there" credo Green Virtue—I'm rescuing "the environment/planet" from humanity!—legitimizes a great many do-nothing environmentalist strategies.

In these next few Reasons, I zero in on Whole Planetude, the 2nd credo—which assures you that all strategies "save the planet" and therefore that anything you do will be effective.

Change a light bulb, save the planet. Recycle a yogurt cup, save the planet. Pick a crisis, any crisis. Pick a strategy, any strategy.

Just do something, anything, any little bit: how many times have you heard *that*?

Whole Planetude: it's a popular American environmentalist logic, which assures you that all "save the planet" actions are additive and interchangeable—and therefore, that no particular action is 100% necessary.

✳ Just add Whole Planetude— what, where, when, who, how much, how fast, *who cares*?

Consider the ultra-popular "save the planet!" listicle: the 12 actions, for example, that you can take to stop climate change in its tracks. Consider too the bargains that Save the Planeteers commonly make with themselves: yes, my Range Rover gets 22 mpg, but hey, I eat all organic and I recycle everything.

As if apples are oranges—but more as if apples are Big Macs are Twinkies are blue-raspberry Slurpees: can you feed your kids anything and have them still grow up healthy?

If you're armed with Whole Planetude, you don't have to advocate for urgent solutions to ultra-urgent crises—climate change, for example, which might eventually half drown New York, Boston, Houston, and Honolulu, along with cities all over the world.

What should we do to reduce CO_2 emissions? Anything!—which allows you to refuse to insist on the most effective strategies.

Where should we reduce CO_2 emissions? Anywhere!—which means you don't have to focus your efforts on the worst actors (oil companies, anyone?), who are actually emitting the most greenhouse gases in particular places.

How much will our existing strategies actually reduce CO_2 emissions, and will it be enough? Just do something, it'll all add up!—which allows you to insist, essentially, that $1 + 1 = 100$.

How do our strategies to reduce CO_2 emissions also contribute to CO_2 emissions? Who cares, something is better than nothing!—which allows you to insist, essentially, that $1 + 1 - 2 = 100$.

✳ Just add Whole Planetude— to Green Virtue

Of course, Green Virtue and Whole Planetude—the dynamic duo of legitimizing "save the planet!" logics—work most powerfully when you combine them.

What do you get, after all, when you add "anything you do will save the planet" to "you're awesome when you save the planet"?

You're awesome for doing anything to save the planet!

> Together, Green Virtue and Whole Planetude powerfully legitimize an "A for effort!," "anything you do is terrific!" approach to climate change and to every other terrifyingly urgent environmental crisis.

Yay for effort, for *climate change*.

And while Whole Planetude blatantly infuses strategies such as pollution trading and offsets—just do something, who cares where—it also double-legitimizes every "sidestep the problem!" strategy that Green Virtue assures you is awesome.

Whole Planetude legitimizes the obsessive Save the Planeteer question "what can *I* do?"[*] Change a light bulb, good for you, you're doing something!

Whole Planetude legitimizes the ever-popular answer "buy

---

[*] see reason #3

something!"* Buy an Energy Star TV, save the planet! Buy 5 TVs, and you'll do 5 times as much to save the planet!

Whole Planetude legitimizes corporate greenwashing: yay, Walmart, every little bit helps!†

And last but never least, Whole Planetude empowers affluent Americans to ignore the essential "who" questions almost entirely.‡ Who is actually responsible for the climate crisis? Humans! How about plastics? Everyone!

You get to avoid so many thorny, difficult questions about whether you, in fact, might generate more CO_2 and toxics and waste than most humans—20 or 50 or 143 times as much?—and about who actually lives with the worst consequences.

✳ Just add Whole Planetude— all of it

Just add up Green Virtue and Whole Planetude—the two "out there" environmentalist credos—and you don't have to substantially, much less massively, challenge our toxic industrial and economic practices.

You don't have to challenge how we, as 21st-century Americans, change environments "in here," mostly vastly badly and inequitably, to create and use the stuff and wealth we rely on daily to live.

* see reason #4

† see reason #5

‡ see reason #6

Just do something, anything, anywhere—any little bit, whatever you can manage. Everyone created these crises, and everyone is suffering the consequences, and everyone will benefit from whatever you choose to do. As long as your SUV is a hybrid, and as long as the box that Amazon uses to deliver your MacBook is recyclable, and as long as Exxon sponsors a few big Earth Day festivals.

A for effort! The Whole Planetude credo infuses and legitimizes a large universe of "just do something" strategies that fail ultimately and egregiously to achieve much of anything at all.

These strategies fail even to begin to tackle the worst messes.

And they dangerously forestall the growing and desperate need to do what's effective.

What, where, who, how, how much, how fast—who cares?

Greenwashing Works, Redux

Protect tomorrow. Today.

—Exxon motto

GREENWASHING, I THINK, MERITS ITS OWN SEP-
arate shout-out as a Whole Planetude strategy.

We're seeing an intensifying whirlwind of greenwashing, after all, as the world's largest, highest-polluting,
and most powerful companies rapidly ramp up their
efforts to convince you they're "saving the planet!"

To fully understand why greenwashing is so effective,
you really have to add Whole Planetude to Green Virtue.

Why *do* Coke and Nestlé expend so much energy and
effort to persuade you that they're leading the charge
on the plastics crisis—we're not plumb daft idiots, are
we?—when they're so obviously leading the charge to
perpetuate the production of single-use plastics.

A for effort! If you're a high-polluting company, you
can invest in any green actions—anything, anywhere,
any little bit—and Save the Planeteers will applaud you,
NGOs will partner with you, and public agencies will
subsidize you.

✳ Just do anything, and we don't care what else you do—or what you fail egregiously to do

Consider, for example, two of the most powerful companies globally—ExxonMobil, the largest publicly traded oil and gas company, and JP Morgan Chase, the largest U.S. bank.

What do these companies do? What else do they do, however, and also what do they fail miserably to do?

Since 2019, Exxon has been buying up solar and wind power in West Texas. Also, Exxon funds the nonprofit Recycling Partnership, founded in 2003—"we're all in this together," the website assures us—and cofounded the nonprofit Alliance to End Plastic Waste in 2019 ("we're all in this together," the website proclaims).

What else? Exxon buys all that renewable energy specifically to power their fast-expanding West Texas oil field operations in the Permian Basin. Also, the company shells out roughly $41 million a year to lobby against climate policies.[*]

In other words, Exxon not only fails to reduce the oil, gas, and plastics production that's driving climate change. They're very actively expanding production, and also using the profits to quash effective action.

Oh, and Exxon and BP, with several other top oil giants,

---

* &, before all this, cofounded the notorious climate-science-denial Global Climate Coalition (1989-2002), which, in 2001, substantially influenced the U.S. refusal to ratify the Kyoto protocol—a moment before the fossil-fuel industry pivoted from "climate change is fake!" to "we're leading the climate fight!"

spend roughly $195 million a year to brand themselves as climate action heroes.

OK, how about JP Morgan Chase? What does the largest U.S. bank do? They tout their switch to 100% renewable energy for their buildings and data centers. Their LEED Platinum HQ boasts 266 bike racks.[*] Also, Chase's CEO voices support for the 2016 Paris climate agreement, and they're facilitating $200 billion in clean energy financing through 2025.

What else? Since the Paris agreement, Chase has lent roughly $65 billion annually to the fossil-fuel industry—36% more overall than any other bank, but 43% more if you only count loans to the 100 companies that are expanding production most aggressively.

Also, Chase has lent more than any other bank specifically to finance the extraction of coal, Arctic oil and gas, fracked oil and gas, tar sands oil, and offshore oil and gas.

And despite a 2020 pledge to wind down financing for coal and Arctic extraction, Chase fails pretty egregiously to do everything that Exxon fails miserably to do.

> The world's highest-polluting companies weaponize Whole Planetude—just do something, anywhere—which empowers them to continue to pollute as much as or more than they have in the past.

* or it did, before Chase razed the 700-ft building in 2020 to build one twice as high

These companies are fiddling while they burn the world—albeit dousing the fires with an occasional bucket of water.

And Save the Planeteers, EPA officials, and Earth Day festival organizers often make it all so easy.

They say, essentially, way to go, Exxon!—you're doing something, it all saves the planet, this is a good start, we're so encouraged that you're trying—instead of saying, hey, Exxon, when, exactly, are you going to finally stop futzing around?

✳ Yes, we care!

How about saying, yes, Exxon, of course we care!—about what else you do and also about what you're failing egregiously to do.

Hey, Exxon, you've known about climate change since at least the late 1970s, right?—and you've also known that plastics are remarkably difficult to recycle. You funded your scientists to conduct some of the early climate research—until the early 1990s, when you decided instead and very strategically to challenge your own results.

Maybe, just maybe, you could show some real, actual, honest-to-goodness goddamn leadership?

You could stop investing your vast resources in CEO and shareholder payouts as your ultimate priority—and start to invest substantially instead in figuring out how to change environments a lot more equitably and sustainably to make, use, and distribute clean energy for all.

You could also stop lobbying, disinforming, tax evading, and greenwashing—all from your shiny new LEED-certified offices and all as if you're trying incredibly hard to ensure there's no tomorrow.

Hey, Exxon and Chase, the U.S. Green Building Council has certified you as LEED (Leadership in Energy & Environmental Design) trailblazers—but seriously?

Yes, you're leading the way—on how to spread disinformation, triple down on high-polluting practices, and ramp up plastics production as plan B, in anticipation of the day when you have to stop turning oil into gasoline.

You're showing us how companies can wring every last dollar out of making our environments ("we're all in this together") increasingly and terrifyingly unlivable.

Hey, let's establish a new green building certification program, OK?—but it'll be exclusively for the construction of corporate buildings and headquarters.

We'll call it Green Roofs for Economic & Environmental Devastation—and your office buildings* can earn a rating of Executive Platinum, Gold Standard, or just plain Certifiable.

What, where, who, when, how much, how fast? Yes, we care!

How about saying *that*?

---

* on your solar-heated campus, where you recycle memos about how to maximize your profits by maximizing social & environmental costs, & where the folks in your top-flight legal, marketing, & accounting departments stash your profits & high-polluting activities offshore, slash your taxes, & lobby against labor & environmental regulations—all while breathing clean air, using low-flush toilets, & enjoying organic açai smoothies from a kitchen with an herb garden, all-natural lighting, & a recycled-bamboo floor

You Wouldn't Create Public Markets to Trade in Cancers, Birth Defects, and Deaths

California [is] a climate leader . . . if we're grading on a curve

— Grist, 2017

CONSIDER CALIFORNIA'S OFT-PRAISED CLIMATE plan—which relies most heavily on carbon trading to achieve over 30% of the greenhouse gas (GHG) reductions.

And which, in 2019, was nicely on track to hit its 2050 targets in 2157—only 107 years behind schedule.

Is there a more "what, where, how much, *who cares*" strategy than the carbon markets for trading and off-sets? You can pay to keep polluting here, if that's easier or more profitable—and we'll ensure that anyone else can profit by polluting less anywhere else.

Wow, what a quintessentially Whole Planetude strategy—boldly, baldly, with no apologies. You do something, anything—somewhere, anywhere—while you also actively and religiously fuel and safeguard the growth-based, profit-maximizing economy that fuels the climate crisis.

These markets are designed so ingeniously to fail, and on such a gargantuan scale, as states, regions, and countries race to establish them and oil and gas companies lobby aggressively for them.

Really, they win the "most terrifying 'sidestep the problem!' strategy" award.*

Trading and offsets, sadly, will never save Miami—for two "um, duh" predictable reasons.

Why, exactly, requires a bit of explanation—but please hang on, as the bizarre complexity of this "save the planet!" strategy makes for a wild ride.

❊ Why trading and offsets programs can't save Miami, "um, duh" reason #1
The programs pump billions of dollars into a growth-based, profit-maximizing economy that is mostly and inherently high-polluting

California, for example, started with this goal: to reduce greenhouse gases while encouraging economic growth. Which is sort of like vowing to win the Tour de France—the

---

* even among all the profit-based "use the problem to solve the problem" strategies—carbon taxes, green-growth consumerism, Cash for Clunkers programs, & on & on

world's most grueling marathon bike race—while funding the opposing teams.

The California Air Resources Board (CARB), accordingly, created a carbon trading program,[*] and here's how it works.

CARB sets a cap on total allowable greenhouse gases and auctions permits to the state's worst polluters—I'll call them SWPs—to emit this total. The SWPs can opt to reduce emissions and sell their permits—if that's more profitable— or they can buy permits to increase emissions, or they can choose to do nothing at all. Banks, hedge funds, and anyone else can register to participate in the auctions, and everyone can sell their permits on the open market.

In other words, trading programs protect polluters' growth and profits with a vengeance. Yes, they incentivize some companies to clean up. And yet, they actively encourage many companies to do absolutely nothing—even when reducing GHG emissions might be incredibly easy or wildly urgent to do.

The programs also dump massive wealth into a profit-maximizing economy, which itself is designed very intentionally to ignore environmental costs. Consider a hedge fund, for example, that buys permits, sells them, and invests the profits in Exxon and other fossil-fuel titans—which then use the money to expand their Gulf Coast operations dramatically and to lobby fiercely against environmental regulations.

---

* in a plan that features other strategies, too—efficiency standards, renewables, recycling, green building standards, forestry projects, & so on

> What carbon markets mostly achieve is to make money fly around—in an economy in which a whole lot of that money will inevitably and actively perpetuate climate change.

Could trading programs possibly be *less* effective? Yes, they can.

While California slightly reduces the cap each year—in theory, to make it increasingly expensive for polluters to emit GHGs—the state is so afraid of hampering growth that they've typically set the cap higher (as in, higher) than the year's total emissions.[*]

Also, the state gives "burdened" industries—including oil and gas companies—their permits for free.

And offsets? Unfortunately, this "let's protect polluters' profits above all" approach applies to these programs as well.

The world's largest GHG offsets program, for example, is the UN's Clean Development Mechanism (CDM)—which the UN launched in 2001 to help the countries that signed the Kyoto protocol meet their reduction targets.

How does the CDM work? The program awards credits to sustainable development projects in "developing" countries, in the Global South and elsewhere—and high-polluting companies in industrialized Western countries can buy credits (in Europe, as part of the EU trading program) to offset their GHG emissions.

In other words, the CDM encourages countries that are

---

[*] & read this again if you want to, but it'll say the same thing

responsible for the highest emissions to continue to emit away. And it subsidizes countries that now emit far less to more fully participate in an economy that's designed to maximize growth and profits more than health and well-being.

Offset schemes, like cap-and-trade programs, actively encourage and subsidize the economic forces that drive climate change.

The CDM, for example, has doled out billions in credits to gargantuan hydropower projects—including notorious new dams in China and Brazil—that displace millions of people, devastate ecosystems, and disrupt viable local economies.

These dam projects, to begin with, emit high levels of methane—a greenhouse gas that is vastly more potent than CO_2.

※ **Why trading and offsets programs can't save Miami, "um, duh" reason #2**
The Whole Planetude approach (who cares what, where, how much) makes the programs incredibly easy to game—and also basically impossible to verify

While carbon markets work to reduce GHGs by using markets that can increase GHGs, the "just do anything anywhere" approach also encourages iffy reductions and outright fraud.

The reductions—if woefully insufficient to begin with—often don't actually happen at all.

You can buy offsets from Carbonfund.org, for example, when you fly from Chicago to Seattle,[*] and the company then pays to plant trees in Panama. Cool, right? And yet, the trees have to grow 30–40 years to reduce CO_2 significantly. They might get chopped down in 15. And planting trees in one spot often just pushes deforestation next door.

Of course, the basic premise of these markets—to ensure that polluters can maximize profits while reducing GHGs—strongly encourages polluters to exploit the Whole Planetude whatever-ness, both legally and illegally.

The CDM, for example, has paid out billions in credits to refrigerant companies in China and India to destroy greenhouse-gas-loaded HFC-23s—which has proved to be a huge incentive to produce more HFC-23s.[†]

The CDM credits go to reforestation projects in Brazil, and loggers promptly destroy the forests.

A great many of the hydropower, low-till agriculture, and other projects—85% to as much as 98%, by some estimates—were exceptionally likely to have happened without any CDM payments at all (despite rules to prevent this).

The recipients of credits engage in funky accounting to qualify—double books, inflating emissions baselines, counting existing national parks as reforestation sites, you name it—and in some cases, companies just take the credits and do nothing at all.

---

[*] you get a certificate!

[†] which the companies can then earn twice the retail price & 75 times the cost of the destruction to destroy—& in 2018, HFC-23s increased at a record pace

STOP SAVING THE PLANET!

And how about trading programs?

In California, companies are allowed to buy state-certified offsets to achieve 8% of their GHG reductions—so basically, up to 8% of the reductions in California might not be happening.

Of course, the polluters in California and elsewhere game the trading markets however they can. They buy up permits, for example, and sell them off at a sizable profit when the cap goes down (despite rules to prevent this).

The best you can say for California's program, really, is that the state seems so far to have avoided the brazen and quite impressive illegality that's plagued the older and larger EU program—and that includes tax fraud, fake credits and accounts, hacking, stealing, money laundering, fake projects, and selling the same credits multiple times.

In sum, WTF?

Or to put it another way, what, exactly, about these "efficient" strategies to tackle the urgent and increasingly apocalyptic climate crisis looks even close to adequate or even remotely effective—or even slightly efficient?

✳ Just double down—on trading and offsets programs that "um, duh" can't save Miami

And yet, in 2019, the public carbon markets alone[*] sent a record $214.5 billion flying around globally—all delaying

---

[*] so add the hundreds of millions of dollars that Carbonfund.org & other private offsets companies pump into the economy

actual, effective solutions and all accomplishing nothing whatsoever to significantly roll back climate change.

Governments and policy makers remain deeply, quasi-religiously committed to an economy that requires unsustainable growth, and to using markets—designed very explicitly to maximize profits—to encourage sustainability and the public good. They refuse to believe that carbon markets just will not work.

Instead, behold the 7 stages of carbon markets: A state or region or country rolls out a trading or offsets program triumphantly, implements it, watches it fail to do much of anything, expresses bewilderment and disappointment, redesigns the program to make it even more market-friendly, renews the program, and announces the renewal triumphantly.

All while Whole Planetude logic—just do anything, anywhere—fuels public support.

In 2017, when California renewed its climate plan to 2030, the Air Resources Board made its trading program even more polluter-friendly.

The new plan adds tax breaks for a lot of manufacturing and agricultural industries. It also extends the free permits, which were originally meant to be temporary, for oil and gas (and other "burdened") companies—which have proceeded to celebrate this measure as one of the industry's biggest anti-climate-policy wins.

Also, it authorizes the state to dip into permit auction fees—which the state is supposed to use to reduce emissions—to backfill this major loss of tax income.

Still not enough? The plan imposes a moratorium on all additional state CO_2 regulations. It also strictly prohibits local and regional CO_2 caps—and did so on the eve of a move by the Bay Area Air Quality Management District to impose one.

All while California issues increasingly higher (as in, higher) numbers of drilling and fracking permits to oil and gas companies.

✳ What trading and offsets programs that "um, duh" can't save Miami do actually achieve

What all this money flying around *can* work to do, sadly, is to increase GHG emissions in low-income neighborhoods—which are predominantly nonwhite, and which are exactly the communities that already suffer devastating levels of sulfur oxides, nitrous oxides, mercury, VOC chemicals, particulates, and other highly toxic co-pollutants.

The carbon markets accomplish this, at least, very efficiently. And this, too, is really just baked-in duh predictable.

Companies can pay to keep polluting here, if that's more profitable—and anyone over there can profit by polluting less? It's not all that difficult to predict that the "heres" will tend to be low-income areas, where the worst polluters have always stashed their pollution so readily and cheaply.

In 2018, more than half the SWPs in the California trading program actually emitted higher GHG levels—yes, as in higher—than when the state launched the program in 2013. By 2019, oil and gas companies, astonishingly, were emitting 3.5% more.

And the increases were happening—we're shocked, shocked—mostly in lower-income areas.

All in all, these stats add up to an exceptionally unhappy "I told you so" moment, as environmental justice advocates object pretty universally to offsets and trading programs.

In California, CARB's own internal environmental justice advisory committee[*] argued vehemently and at length that the climate plan should under no circumstances rely on carbon trading—and then submitted a 40-page official objection to the final plan.

In 2017, CARB ignored vehement pleas once again and chose to double down on its trading program.

Or here's the no-sugar way to put it. The state of California stubbornly maintains a place-blind market to trade in greenhouse gases, which, while doing essentially nothing to roll back climate change, encourages people to profit by trading in cancers, birth defects, and deaths.

* with a quondam website page on which CARB featured the "whole Earth" image

 **Reason**

Everyone Hates Environmentalists (Now Add Whole Planetude)

The earth's most vulnerable populations experience the greatest suffering from environmental degradation
> —Communities for a Better Environment, Southeast L.A.

The air sucks anyway
> —spectator at a "rolling coal" contest, McHenry County Fair in Illinois, 2016

WHO CARES WHERE? SERIOUSLY?

If you live in Southeast L.A., you might care quite a lot when California doubles down on their Whole Planetude "who cares where" climate strategy.

We already stash the lion's share of the disastrous co-pollutants in your neighborhoods. A lot of knowledgeable people (including from Southeast L.A.) actually warned, with everything but a Bat-Signal, that toxic emissions in your community would likely go up. And they were right.

How would *you* respond to the Air Resources Board issuing free permits to pollute? Or to hedge funds reaping huge profits by speculating on those permits to pollute?

How would you respond to all the tree planting in Panama? Or to the 5 largest U.S. oil companies spending $195 million annually to market themselves as clean energy heroes?

Maybe you live amidst the 150 petrochemical plants in Cancer Alley, in southeastern Louisiana—an 85-mile, heavily African American corridor between Baton Rouge and New Orleans. St. James Parish alone hosts 32 plants, and the cancer risk in St. John Parish is 50 times the national average.

Wouldn't *you* hate all this "just do anything, anywhere," Whole Planetude–besotted nonsense? Maybe you'd cringe a little when you hear all this concern about "people and the planet"?

Of course, while we stash all this industrial pollution importantly and predominantly in areas where Americans of color live and work, a huge abundance of lower-income white communities are likewise besieged.

So maybe you're one of the few people left in Hinkley, California—the majority white *Erin Brockovich* town—where the utility company PG&E is buying out residents, after dumping 370 million gallons of wastewater, laced heavily with the potent carcinogen chromium-6, into the groundwater. The plume continues to spread through the aquifer, despite ongoing cleanup efforts.

Would you feel encouraged by PG&E's electric hybrid trucks? Would you care—at all—about JP Morgan Chase's vow to power their operations with 100% renewable energy?

You might live in the diverse but majority white and widely impoverished Appalachian coal region—long dubbed a "national sacrifice zone"—where advanced black lung cases have been surging, and where strip-mining has blown away the mountaintops, obliterated waterways, and contaminated your groundwater.

Maybe you live in Albuquerque's mostly Latino South Valley area, which hosts the region's landfills, sewage treatment plants, widely leaking underground oil and gas storage facilities, and over 90% of the area's EPA-designated toxic release sites.

You might live in a South Side Chicago neighborhood—in the heavily African American Roseland area, or in the mostly Latino Little Village—where the expansion of high-polluting industries proceeds in tandem with the expansion of cleanups and gentrification on the more affluent and more heavily white North Side.

If you live in Flint, Michigan, or Newark, New Jersey, you have to cope with devastating air pollution and also with dangerously high levels of lead in the tap water.

Maybe you live on the Crow Reservation in South Central Montana, where the rural wells and municipal water systems alike tend to be heavily contaminated.

Wouldn't you care enormously about what, where, who, how much, and how fast?

Of course, I couldn't disagree more vehemently with many Americans—mostly white—who, while struggling so precariously to survive in these communities, blame environmentalists more than Exxon and Walmart and Chase.*

Or who spout racist invective to justify their own precarity. Or who vote for ultraright candidates whose policies will only perpetuate both the drastically inequitable economy and the mega-pollution it inevitably creates.

And. Yet. But. Still.

Why do so many Americans who suffer the most from environmental crises—and who benefit the least from creating and perpetuating these crises—hate environmentalists?

Why, whether you live in St. James Parish in Louisiana's Cancer Alley or in a coal town in Kentucky, might you believe not only that "saving the planet!" is wholly and nonsensically irrelevant, but also that Save the Planeteers, corporate behemoths, and public agencies alike are actively betraying your interests?

Maybe you don't trust rich people, or perhaps you don't trust white people—or maybe history has given you plenty of reasons not to trust white affluent do-gooders.

Maybe you just don't trust government.

And maybe it's because "save the planet!" environmentalism offers you very few reasons to believe otherwise.

* the "rolling coal" fad—belching black clouds of smoke by souping up engines & removing emissions controls—being one of the most creative & fascinating ways that this group of environmentalist haters performs their hatred. A 2nd way: "death before Prius" T-shirts.

You Can't Tackle How We Are All in This Together Unless You Tackle How We Are *Not* All in This Together

We are running out of time, and our governments have failed to act. . . . The time for denial is over

—Extinction Rebellion

WHAT IF WE FOCUS OUR EFFORTS ON THE WORST environmental messes[*]—which we've long tended to stash in low-income areas?

How about a geysers-erupting-up environmentalism?

I think this might just be a brilliant idea, for two big reasons.

One, the geography of these messes is unfair—just wildly and democracy-smashingly unconscionable. Your access to clean air and water, healthy food, parks and greenery, and so on should not depend on how much money you have.

---

[*] say, the billions of tons of petrochemical waste in Cancer Alley in southern Louisiana

And two, we don't have time for all this "saving the planet"—for all this tiresome Green Virtue and all this pointless Whole Planetude, which just allows you to assure yourself that all these "A for effort!" do-nothing strategies are doing something.

The Antarctic ice sheets are collapsing: we're talking a 10-foot sea-level rise. The great majority of U.S. municipal water systems—your zip code be damned—are now loaded with the highly toxic "forever chemicals" PFAS.

The Sixth Extinction is happening around you. You're literally eating a heaping plate of microplastics each year. I could go on and on.

✳ We need a Green New Deal—or call it a Great Changes Environmentalism

Yes, yes, we need a Green New Deal, or call it the Geysers-Erupting-Up Plan to Battle Environmental Crises.

We need a massive, transformative initiative to create a vastly greener and more equitable economy—as a growing army of people are demanding, in the U.S. and around the world—and to use it to fight climate change, banish poverty, and clean up the worst messes.

However: the 21st-century New Deal *has* to be Green, not because we need to achieve two urgent but essentially *different* goals—to democratize the economy and also to fight climate change.

> A new New Deal must be Green because
> the *foundation of a democratic economy*
> is to change environments as sustainably
> and equitably as you possibly can.

This Great Changes environmentalism insists that we change environments fairly and well to live, work, and play.

And it insists that clean air and water, healthy nonpoisonous food, and thriving ecosystems are foundational civil rights.

✳ Call it the Laws & Taxes & Jobs R Us New Deal

Or call it the Government Is Us Economy—but let's insist that the most basic, ABC purpose of jobs, businesses, and government should *not* be to change environments phenomenally badly and inequitably so that a few people can enjoy ever-increasing wealth.

No, the ABC purpose should be to change environments fairly and in ecologically healthy ways—this might just be a brilliant idea—to provide and/or engage in energy, water, food, infrastructure, housing, manufacturing, retailing, health care, transportation, education, arts, communication, research, defense, recreation, and you name it.[*]

* by means of fair taxation, public ownership of basic infrastructure, a GDP that values health & well-being above all . . . &, well, please see the 39 Ways to Stop Saving the Planet

Call it the Toxics Be Gone New Deal—in which our laws encourage and mandate the poison-free creation of our stuff and wealth and our taxes fund Departments of Green Chemistry, Regenerative Agriculture & Forestry, Clean Manufacturing, and Cleaning It All the Hell Up.

We could call it the Shouldn't Strawberries Be a Lot More about Strawberries Plan—in which a company's sole motive to grow strawberries should not be to maximize profits, and therefore to use minimum-wage labor and abundant poisons to grow as many tasteless fruits as possible that a global corporation can ship as far as possible. Rather—how brilliant is this?—the primary goal of an agricultural business should be to grow healthy and scrumptious food, for a reasonable profit, in healthy soils in the midst of thriving ecosystems.

Call it the War on Greed and Pollution, in which the goal of a health care business—and this is definitely brilliant—should be to provide excellent health care instead of to maximize profits by restricting access to excellent health care. A school should be about education, and making chairs should be about making great chairs.

Cleaning up a disastrous oil spill should be—first and most passionately—about thoroughly cleaning up a disastrous oil spill.

Call it the Stopping Climate Change—or COVID-19, or Any Other Lethal Global Crisis—Should Be about Stopping a Lethal Global Crisis (& Shouldn't That Be a No-Brainer) Initiative.

You could call it the Huge Fat Initiative to Massively

Rethink a T-Shirt—so that making, transporting, buying, and discarding your tween daughter's soccer T-shirt requires nontoxic agriculture and manufacturing, public utilities and clean renewable energy, well-paid jobs, safe working conditions, healthy and diverse ecosystems, fair taxes, cooperative ownership, well-funded education and research, laws that keep companies accountable, short transport routes wherever possible, a deep interest by all involved in the T-shirt's quality and durability, and essentially zero throwing of anything away.

Call it the You Shouldn't Have No Choice but to Work in Jobs That Are Ruinous to People and Environments Great New Plan.

Or call it a Great Changes Environmentalism, or a Green New Deal—and regardless, we definitely don't have time for all this "what can *I* do?," "buy something!," carbon speculating, green growth, greenwashing, just do something, A for effort!, self-deluding, trickle-down "save the planet!" utter ridiculous bullshit.

Call it a Good Intentions Aren't Anywhere Close to Good Enough Environmentalism.

We're going backward on climate change, plastics, groundwater pollution, and almost every other environmental crisis.

We're going backward—maybe more slowly than before, and maybe not.

We're not even beginning to make enough of a difference, and everyone, ultimately, is going to lose—and in many ways that you and I and our many public officials might still be deeply loath to even begin to acknowledge and imagine.

You can call it the 21st-Century Green Initiative to Finally Stop Futzing Around Already.

Call it the An Equitable Economy Is Not a Radical Idea New Deal.

Or just call it the Mobilization against Insanity.

WAYS TO STOP SAVING THE PLANET

A Stout but Unexhaustive List of Things That You Can Personally Do

Kinds of things

 SOCIAL CHANGE — make change together

 HOW THE WORLD MAKES MONEY GO 'ROUND — change how *you* do it

 WORD UP — talk differently, think differently

 STUFF MATTERS — change your relationship with your stuff

 CIVIC INSTINCTS — be a citizen

 ECO INSTINCTS — see and enrich your ecologies

 YOU KNOW — learn more

 BAMBOOZLE-FREE ZONE — insist on the larger context

 SPECIAL POWERS — arm yourself with indispensable methods

 OF COURSE — just do it

 OF COURSE NOT — just don't

 GOLDEN RULE — Golden Rule

 ★ = **ESPECIALLY IMPORTANT WAYS**

★ **Pay more attention to how you make money than to what you use it to buy**

Ideally: change environments as sustainably & fairly as you can by how you work, bank, &/or invest.

Instead of: change environments wildly unsustainably & unfairly by how you work, bank, & invest—& then use your money to buy a Tesla.

SCRIBBLE ZONE
Write, draw, ponder . . .

② Have a "where are the worst messes" join-up, pt 1—where are they?

SOCIAL CHANGE

Where are the worst messes—& how far from where you live & work?

Where are they in your neighborhood, town, county, state—& then, your region, country, continent, & world?

YOU KNOW

Think: air & water toxics, dumps (legal & illegal), landfills, Superfund sites, other industrial sites, soil contamination, desertification, & what else.

Useful starting point: EPA's Superfund list.

SCRIBBLE ZONE
Write, draw, ponder . . .

★ Redefine environment

Good way to start: talk & think about environments instead of "the environment."

Also try this: think about how a tree or table is "environmental," instead of asking whether it's environmental or not.

And always keep track of: the essential environmental-ness of Yosemite, the Everglades, trees, tables, refrigerators, plastic toys, chicken fingers, city streets, the internet, & basically everything else.

3

Say What?

WORD UP

SCRIBBLE ZONE
Write, draw, ponder . . .

**STUFF
MATTERS**

★ Quality vs quantity, less is more

If you're a producer: make better stuff, sell less stuff, & spread the $$ fairly.

And/or if you're a consumer: buy better stuff, buy less stuff , & you'll likely come out ahead—as a high-quality mini-vac might last longer than 3 or 4 cheap crappy ones.

Unfortunately, in a "cheap replaceable crap" economy: if you can't afford the better stuff, you might ultimately have to pay more to vacuum your living room.

SCRIBBLE ZONE
Write, draw, ponder . . .

Redefine ambitious

Yes, ambitious!: our climate plan will phase out fossil fuels by 2030, or we're so screwed—& hey, is 2028 possible?

Not: our climate plan will reduce fossil-fuel emissions significantly by 2050—& Exxon signed on!

Also redefine: cutting-edge, pioneering, aggressive.

SCRIBBLE ZONE
Write, draw, ponder . . .

**STUFF
MATTERS**

How green is your building?—inside & out

Good rule of thumb: pay at least as much attention to how you live & work inside buildings as to the buildings themselves.

A certifiably green building: you earn, spend, & invest as sustainably as possible inside it—& then you worry about dual flush toilets.

Just certifiable: you install dual-flush toilets & state-of-the-art air & water filters—but you earn, spend, & invest in ways that generate dangerous air & water pollution around the world.

SCRIBBLE ZONE
  Write, draw, ponder . . .

★ Redefine economy

WORD UP

What an economy is (suggested definition): how, as a society, we change environments to create stuff & wealth to provide people's needs & wants, & how we distribute the benefits (stuff & wealth) & the pollution & other costs.

What an economy is a lot more than (dictionary definitions): "a network of producers & consumers," "how money is made," "wealth from business & industry," "how people allocate their limited resources to satisfy their nearly unlimited wants."

What an economy includes: paid work, unpaid domestic work (childcare, cleaning the bathroom, baking bread, walking the dog, painting the kitchen, etc.), volunteer work, bartering, gardening, & what else.

What an economy should do: sustain the health & well-being of people & environments—which is actually what most of us care most about—by changing environments fairly & sustainably.

Also redefine: value, costs, GDP, efficiency.

And finally: wouldn't it be nice if we didn't need special laws & guidelines that "allow" a company to focus not solely on profits but also on the health & well-being of people & environments?

SCRIBBLE ZONE
Write, draw, ponder . . .

Join up locally—neighborhood environments

Share & create habitats & sustenance for all: shared gardens, school farms, green playgrounds, urban forests, greywater systems, bird-friendly landscaping, stormwater basin mini-parks, butterfly gardens, bee-friendly planting, compost depots, herb gardens, public orchards, front-yard edibles, nature playscapes, greenbelts, & what else.

SOCIAL
CHANGE

SCRIBBLE ZONE
Write, draw, ponder . . .

STUFF
MATTERS

Love the stuff you're with

'Cause the greenest car, toaster, or vacuum cleaner is most often (if not always) the one you already have.

SCRIBBLE ZONE
Write, draw, ponder . . .

Vote

Vote in all public elections: for U.S. president, state senate, judgeships, city council, zoning commission, board of education, & what else.

What to do in advance: homework!—especially for Propositions A through Q (many of these could save or kill you), & see what info you can find on the 530 judges (ditto).

Also: why isn't Election Day a national holiday?

CIVIC
INSTINCTS

SCRIBBLE ZONE

Write, draw, ponder . . .

Have a "where are the worst messes" join-up, pt 2—see for yourselves

SOCIAL
CHANGE

Go, find, see: in your backyard, neighborhood, city, town, county, state.

Helpful to give, helpful to take: "toxic tours." Check out, as examples, the tours by Communities for a Better Environment (Southeast L.A.), Texas Environmental Justice Advocacy Services (Houston), Little Village Environmental Justice Organization (Chicago).

YOU KNOW

SCRIBBLE ZONE
Write, draw, ponder . . .

Stock up on "yes, it's possible, it's happening as we speak" models

Just a few of the things that are happening, in just a few of the places (& no, not just in Europe):

* single-use plastics bans (EU, Mumbai)

* worker co-ops (everywhere)

* worker ownership (yes, U.S.)

* citywide green infrastructure (Copenhagen)

* regenerative agriculture (everywhere)

* affordable health care for all (every non-U.S. affluent country, just to start with)

* green chemistry (in your supermarket)

* 32-hour workweek (Netherlands), 4-week-vacation policies (lots of places)

* bans on toxics in cosmetics & cleaners (EU)

* bans on ultra-toxic pesticides (EU)

* only glass beverage bottles allowed (Denmark, with a near 100% return rate)

* 95% food waste recycling (S. Korea)

* right to repair laws (EU, Massachusetts)

* sustainable stormwater management (Netherlands, lots of places)

* public banking (lots in U.S.)

* land trusts (lots in U.S.)

* well-being vs GDP as economic health index (New Zealand)

Great umbrella sites to start with: Beautiful Solutions, Story of Solutions, New Economy Coalition blog.

SCRIBBLE ZONE
Write, draw, ponder . . .

★ Pay a lot more attention to how you make money than to how you give it away

Ideally: change environments as sustainably & fairly as you can by how you work, bank, &/or invest.

Why bother: work in a high-paying job for a billion-dollar global company that devastates people & environments, minimizes taxes, & invests its profits & retirement funds in billion-dollar global companies that do the same—but which also uses 2% of its profits to create a nonprofit foundation, which invests 98% of the 2% in the same companies, & which then enjoys deep tax cuts to spend or give away the 2% of this 2% to tackle the social & environmental problems your company & its foundation create by how they earn, spend, bank, & invest.

14

OF COURSE

Give some money away, & only if you can—$10 or $10 billion

If you have $10 billion to give away: maybe pay a lot more attention to how you work, bank, & invest.

SCRIBBLE ZONE
Write, draw, ponder . . .

★ Knowledge is power, right?

YOU KNOW

A few of my own favorite go-tos on environment ⇆ economy:

Websites: New Economy Coalition, Climate Justice Alliance, Story of Stuff, Democracy Collaborative (quick hack: their "What Then Can I Do?" working paper).

Media: *Guardian* (George Monbiot's columns & everything else), ProPublica, Grist, Civil Eats.

Books: Anand Giridharadas's *Winners Take All*, Leah Hager Cohen's *Glass, Paper, Beans*, David Klein & Stephanie McMillan's illustrated *Capitalism & Climate Change: The Science & Politics of Global Warming*.

Legislative documents: why not read the Green New Deal House bill?

A fave environment ⇆ economy thinker: Naomi Klein—*This Changes Everything* (book, website, everything), *On Fire* essays, Intercept essays & videos, *Nation* essay "Capitalism vs. the Climate" as a quick hack into her central ideas.

A "what is environment" classic: William Cronon's essay "The Trouble with Wilderness"—& you could check out my own essay "Thirteen Ways of Seeing Nature in L.A."

A great "what is economy" manifesto: Ethan Miller's "Occupy! Connect! Create!" in the wonderful artist-fueled *Guidebook of Alternative Nows* anthology.

SCRIBBLE ZONE
Write, draw, ponder . . .

Know your ecologies—birds & insects & trees, oh my!

Why be a nature geek: if you wanna create more sustainable worlds, you gotta know a lot about who & what we cocreate our worlds with.

How to geek out: watch for wild things, know them, create cohabitats for them, slow down to see them—& find & follow around your local eco-geeks.

How I geek out: I ♥ birds—I mean, just google blackburnian warbler—but also, birds show me the plants they rely & hang out on, the insects & fruits they eat, the creatures who eat them, the rocks & soils that support it all, & the changing climates it all happens in. They show me how our worlds work & how we don't even begin to create our worlds by ourselves.

Inspiration: Community Nature Connection, Black Birders Week, Outdoor Afro, Latino Outdoors, City Nature Challenge.

16

ECO
INSTINCTS

YOU KNOW

SPECIAL
POWERS

17 SHTEAMS—ways of knowing & solving

Argue for the importance of: science, humanities, tech, engineering, art & design, math, social science.

Why do this passionately: we can't possibly solve the climate or plastics or any other urgent house-burning-down crisis without all these ways of knowing & solving.

Why reinvent the wheel?: be sure to include traditional, local, & indigenous knowledge—about health, ecologies, agriculture, & so on—that people have developed & refined for a century or far longer.

Reject, of course: denial of these essential ways of knowing.

Be sure to reject: the ever-burgeoning humanities denial—unless you're sure we can dig our way out of climate change with little understanding of ethics, inequities, ambiguity, why different people think differently, & generally why people do what they do & think what they think (including why so many people deny the importance of science or humanities).

★ Environmentalist Golden Rule: be an IMBY

GOLDEN RULE

The Great Changes environmentalist's motto: "Don't help stash industrial toxics & waste in other people's neighborhoods that you wouldn't have them stash in yours."

Producers, consumers, policy makers: don't do it, don't help companies do it, don't subsidize it.

SCRIBBLE ZONE
Write, draw, ponder . . .

Redefine extremist

Extremist?: our government & economy should work to maximize health & well-being—so that everyone has access to the stuff & wealth they need to thrive.

Or is this?: our government & economy should work to maximize profits & growth—to make a few people spectacularly wealthy & to impoverish the majority.

Also redefine: crazy, radical, fringe, impractical, impossible.

Words to live by, from Winona LaDuke: "someone needs to explain to me why wanting clean drinking water makes you an activist, & why proposing to destroy water with chemical warfare doesn't make a corporation a terrorist."

Creative is as creative does

SPECIAL
POWERS

You can't overstate: the power of the arts to see, know, reimagine, communicate, challenge deep-rooted assumptions, make visible what's invisible, & actively make change happen—so read, look, watch, listen, & go creative, solo & with others.

How to reinvent the art museum: MoMA reimagined in Claire Bishop & Nikki Columbus's essay "Free Your Mind."

A shout-out to climate fiction: & for suggestions, you can check out the Burning Worlds column in the *Chicago Review of Books*.

A shout-out to participatory projects that shout the "public" in public art: e.g., check out the umbrella sites Creative Time, U.S. Department of Arts and Culture (toolkits included), CURRENT: LA, A Blade of Grass.

21

SOCIAL
CHANGE

★ **Have a "where are the worst messes" join-up, pt 3—with people trying to clean them up**

Join up with others—with the people who live in or closest to these sites & with any & all allies—to fight to clean it all up in your neighborhood, town, county, state, & beyond.

Useful website: Climate Justice Alliance, to find groups & to volunteer.

SCRIBBLE ZONE
Write, draw, ponder . . .

Join up locally—government & economy R us

Share & create democracy: food co-ops, public & community banking, credit unions, lending circles, electricity co-ops, public policy advocacy, community land trusts, local business support, all sorts of worker co-ops, rural ride sharing, agricultural co-ops, tenant associations, high voter turnout, backing candidates for public office (rock the planning commissions), farm shares, & what else.

SOCIAL
CHANGE

Starter toolkit: New Economy Coalition's Pathways to a People's Economy.

SCRIBBLE ZONE

Write, draw, ponder . . .

Redefine renewal, development, & revitalization (it's the people, dummy)

WORD UP

Yes, this is urban renewal!: invest in the residents of low-income neighborhoods—by drastically improving access to parks & greenery, public services, good jobs, livable housing, healthy food, clean air & water, high-quality health care, cultural centers, high-quality public education, & what else.

No, just urban displacement: clean & green up low-income neighborhoods, & then allow soaring real estate values to drive out the residents—generally Americans of color—amidst the arrival of expensive condos, pricey galleries, & a Cafe Gratitude.

Yes, this is rural economic development: invest in struggling communities by facilitating transitions to regenerative agriculture & clean energy—with farm co-ops, job training, strong farm-to-city networks, fair trade partnerships, & what else.

No, just f**king things up: generate poorly paid jobs & crazy pollution, & often drive people from their communities—with fracking, for example, or with huge-corporation agriculture or massive new warehouse developments . . . or with massive prison complexes, to house & make money off of mostly Americans of color from the low-income communities that cities consistently & egregiously fail to invest in.

Inspiration from my hometown St. Louis: Walter Johnson's book *The Broken Heart of America*.

SCRIBBLE ZONE
Write, draw, ponder . . .

24 Join up regionally, nationally, globally

SOCIAL
CHANGE

Shout-out to big global environment ⇆ economy campaigns: Extinction Rebellion, Sunrise Movement, 350.org, Youth Climate Strike, & Zero Hour, to begin with.

Brought to you by: younger people, mostly, who are tired of waiting & who deserve a very different future than the one they increasingly fear they are about to inherit.

SCRIBBLE ZONE
Write, draw, ponder . . .

Join up locally—share, trade, & repair

All the rage: tool libraries, repair cafés, sharing depots, libraries of things, bike repair co-ops, fashion swaps.

STUFF
MATTERS

Your broken vacuum cleaner: now works like a dream. Your favorite coat: 3 more winters. Who doesn't want a chocolate fountain?—& do you really need your own lawn mower, card tables & folding chairs, pressure washer, & popover pans?

A special shout-out: food & garden-bounty exchanges.

SOCIAL
CHANGE

Inspiration for a great 1-day event: the Fallen Fruit collective's public fruit jams.

SCRIBBLE ZONE
Write, draw, ponder . . .

26 Have a "how do we do a lot of stuff without buying a whole lot of stuff" join-up

STUFF MATTERS

How can you . . .

a) have fun

b) express yourself

c) celebrate

d) give gifts

e) show love & gratitude & appreciation

SOCIAL CHANGE

. . . without buying tons of stuff & especially tons of cheap throwaway crap?

Icebreaker: pick a fave from Grist's list of 79 climate-friendly gifts.

SCRIBBLE ZONE
Write, draw, ponder . . .

Go ahead & buy some stuff—new & used

OF COURSE

Sure, buy some stuff you need & want—just not shopaholic-style, & ideally in conjunction with sharing/trading/repairing, & with a severe-ish allergy to cheap single-use throwaway crap, & not mostly when you're just really bored.

Ideally: from stuff-makers & stuff-sellers whose primary motive is to make & sell good stuff at a reasonable price.

Also, maybe: have a "what do we really need & want & why" join-up.

Inspiration: *Portlandia*'s "Instant Garbage" sketch.

SCRIBBLE ZONE
Write, draw, ponder . . .

**CIVIC
INSTINCTS**

28

Taxes R Us

Pay taxes: no cheating. Seriously, no cheating, & here's a "working on my federal taxes" 2-step visualization exercise.

Step 1: take 5 deep breaths, & imagine that you're placing all your negative thoughts—e.g., about fossil-fuel subsidies, endless wars in the Middle East, the initial COVID-19 pandemic response, the fact that Exxon & Amazon might pay less than you do this year—inside a big, big box.

Step 2: take 5 more deep breaths (& maybe a pot brownie), & imagine your $$ streaming around the world & turning into parks, clean air & water, Medicare, small business loans, mail, schools, Pell Grants, NPR, Social Security, bike lanes, the CDC, the NEA, global disaster relief, NIH-funded research, & thousands of things you may love, support, or rely on.

Live with wild things—home ecologies

ECO
INSTINCTS

Share your home with: dogs, cats, birds, violets, lemon trees, & just basically lots of wild things.

You take care of them, & they take care of you: they remind you daily that we cocreate our worlds, & they show you how to do it.

You don't have to share your home with: roaches, rats, & aphids—especially if your wild-thing housemates like to eat them.

Personal shout-out to: Samantha, Solomon, Rookie, Shiloh, Dante, Lola, & mountain goat Charlotte.

SCRIBBLE ZONE
Write, draw, ponder . . .

**BAMBOOZLE-
FREE ZONE**

Recalculate costs—yours, mine, ours

Count *all* the costs—& ask, who pays them?

Case study A: "it's cheaper on Amazon"—but not really, if you count *all* the costs, of fossil-fuel emissions, & of clogged streets, & of landfills' worth of throwaway packaging, & of a huge number of people working in unsafe conditions for low wages & no benefits, & of the destruction of local (& nonlocal) businesses & revenue, & of local tax breaks that rob your public coffers, & of Amazon paying zero to minimal federal taxes (a $29 million refund on $11.2 billion in income in 2018), &, of course, of the inevitable price hikes once all the competitors drop out . . . & of what else.

Case study B: "it's cheaper to replace this stupid piece of crap mini-vacuum than to fix it!"—but not really, if you count *all* the costs, of pollution & waste & horrible labor practices, to make as much stuff as possible as cheaply as possible (not least so you'll say, "this stupid piece of crap mini-vacuum, it's cheaper to buy a new one!").

All in all: the less affluent you are, the more you pay for the cheap mini-vac on Amazon—but everyone pays hugely more than the $24.99 retail cost.

Tell a frickin' joke

Seriously: funny can be a lot more powerful than yelling & lecturing—especially when the other side is screaming & ranting.

The powers of funny: to break down people's defenses, expose hypocrisy, expose the absurdity of ideas that people might take for granted, & what else.

See, for a textbook example: how the town of Wunsiedel, Germany, effectively turned a neo-Nazi march into an anti-Nazi walkathon.

A few E-faves, for news analysis with a light chuckle: Grist, The Story of Stuff.

A few E-faves, for nailing hypocrisy:
"9 Rules for the Black Birdwatcher"; *The Onion* (start with "New Prius"), The Yes Men ("Black Rock," "Coal for the Rich"), *Portlandia* ("Colin the Chicken"), the graphic novel *As the World Burns: 50 Simple Things You Can Do to Stay in Denial*.

BAMBOOZLE-
FREE ZONE

Just say no to blaming individual choices, instead of root causes, for huge systemic problems

Here's a big example (sound familiar?): our public policies deny Americans of color access, decade after decade, to public services, good jobs, homeownership, healthy food, good education, clean environments—& then you blame unemployment, poor health, & rundown neighborhoods on the residents of color to whom our policies have consistently denied access to all those things.

Think bigger & resist!—because blaming individuals more than racism or an inequitable economy is a great American tradition.

Also just say no to: hailing a billionaire to the skies for his/her great individual choices—instead of asking how tax, inheritance, corporate-subsidy, & other hugely inequitable public policies in the U.S. add up to welfare for wealthy folks.

132

39 no-brainer things

OF COURSE

Use recycled paper, avoid bottled water, use efficient light bulbs (after the old ones burn out), turn the lights off when you leave a room, use nontoxic cleaners, walk & bike, use public transit, carpool, recycle (no wish-cycling everything including the kitchen sink), eat less meat, use nontoxic paints & floor finishes etc., air-dry laundry, go easy on paper towels, carry a reusable water bottle &/or coffee mug, raise the AC, lower the heat, reuse boxes & bags & other "single-use" containers, minimize food waste, use nontoxic shampoo & cosmetics etc., print on both sides, go solar if you can, turn off devices overnight (!), eat organic if you can, use nontoxic pet products, BYO takeout containers, garden organically, use reusable shopping bags, compost, use a low-flow showerhead, go for nontoxic landscaping, try less lawn & more habitat, turn off power strips when you're not using them, fill up the dishwasher before you run it, use wax & other reusable wrappers, swear off plastic plates & cups & utensils, don't leave the water running, eat sustainably caught fish, don't litter, avoid plastic food packaging.

Bonus thing: read MacKay Jenkins's *ContamiNation.*

OF COURSE NOT

Ultra-virtuous sacrifices you shouldn't make

2 reasons why: they'll achieve nothing to refreeze the glaciers, while very successfully making the people you love exceptionally unhappy.

So just don't (real examples): refuse to visit your mother if she lives more than 100 miles away, break up with your girlfriend whom you truly love for the sole reason that she lives 4 states away so omg the carbon footprint, refuse to end a bad marriage 'cause 2 households use more energy, refuse to let your kids join the local baseball league 'cause it's 18 miles away, or refuse to have kids in the first place if you want them.

SCRIBBLE ZONE
Write, draw, ponder . . .

Just say no to solutions that only treat the symptoms of huge systemic problems—& that ignore or actively perpetuate the root causes

One sample set of bogus solutions: the "tech will save us!" romance.

Example: GMOs to make crops pesticide-resistant.

Why it won't work: sure, you'll increase short-term yields, but the big GMO companies very intentionally use this technology & the patents to perpetuate toxic & inequitable corporate agricultural practices—which maximize profits instead of nontoxic food, healthy ecologies, fair & safe working conditions, & what else.

Another set of bogus solutions: the siren call of "convenience."

Example: ordering most of your dinners online to save time.

Why it won't work: sure, you'll save a little time tonight, but why on earth don't you have 40 minutes to make dinner, go to a store, or do anything else?—& why do we put up with 12-hour workdays, long commutes, poor public transit, lack of affordable childcare, 7 days' vacation in a year, & what else.

Insist on big solutions!—because yes, ignoring the root causes of big problems is a great American tradition.

SCRIBBLE ZONE
Write, draw, ponder . . .

Have a "figure out where our water & gas & electricity come from, & where our trash & recyclables go, & where the water in our faucets & toilets goes" join-up

36

ECO
INSTINCTS

Useful way to begin: google up a storm, & also call & email your public works departments, water companies, waste transfer centers, & so on.

Most of us don't ask—& it's really not that hard to find out.

SOCIAL
CHANGE

SCRIBBLE ZONE
Write, draw, ponder . . .

**BAMBOOZLE-
FREE ZONE**

Just say no to greenwashing

Ask, if a Coke ad pops up on your Twitter feed urging you to recycle: how might this "thanks for recycling!" ad allow Coke to continue to produce 3 million metric tons of plastic annually?

Ask, if your "green business" class in middle school, high school, college, or grad school cites Exxon as a model: how do Exxon's green acts work to legitimize their wildly un-green practices—& does it all add up to a net positive?

SCRIBBLE ZONE
Write, draw, ponder . . .

We ♥ the public sphere—share, fund, enjoy, expand, fight for, & appreciate it

CIVIC
INSTINCTS

Yours, mine, & ours yet?-or just a few things everyone should have access to: parks, postal service, transit, beaches, clean air, clean drinking water, education, rec centers, clean & affordable energy, streams and rivers, libraries, childcare, art & cultural venues, zoos, affordable health care, swimming pools, basketball courts, the Internet, & what else.

Also should be public (even if they're less lovable): national defense, criminal justice, sewage management & all other basic infrastructure, & what else.

SCRIBBLE ZONE
Write, draw, ponder . . .

39

Have a "how does my cell phone connect me to people & environments—but not through my contacts list" join-up

ECO
INSTINCTS

Or you could focus on your freezer, your favorite T-shirt, or your Spotify list: how does this thing—that I use every day, or that I bought at a concert, or that I dance to in the kitchen— connect me to people & environments around the world?

SOCIAL
CHANGE

So this doesn't become a 3-year join-up: make it a game, with teams & 10-minute rounds, to come up with the most connections.

Bonus round: how does a visit to your favorite wild spot—a local park, an abandoned field in the neighborhood, a riverside wildlife refuge, an alpine valley you hike 3 days to get to in a national park—require & connect you to people & environments around the world?

Acknowledgments

Huge thanks to my angel investors—the Laszlo N. Tauber Family Foundation, Princeton Environmental Institute, Rachel Carson Center for Environment & Society, Princeton American Studies department, and UCLA Center for the Study of Women for generous long-term fellowships; and Grey Towers National Historic Site, the Women's International Study Center, the Bill Lane Center for the American West at Stanford University, and Aspen Words with Isa Catto and Daniel Shaw for essential short-term hideaways.

Great thanks to the Washington University–St. Louis folks who, for four years and counting, have ensured I have a rich intellectual home and an actual office—Bruce Lindsey, Heather Woofter, and Carmon Colangelo at the Sam Fox School of Design & Visual Arts, and also Liz Kramer, Jesse Vogler, and the lovely "office as art" Eric Ellingsen. Thanks, too, to my WUSTL humanities, environmental studies, and larger Sam Fox communities, and to an abundance of beloved former colleagues at Princeton University and at the Rachel Carson Center, with special gratitude to Bill Gleason and Christof Mauch.

Great thanks, too, to my many St. Louis comrades who

help me see and think about "environment + justice" via my complicated hometown—with special props to artist/curator extraordinaire Gavin Kroeber, Allana Ross, Jennifer Colten, Aaron Owens, Angela Miller, Andrew Hurley, and Flannery Burke. Abundant thanks to innumerable colleagues across the country who invited me to their campuses to try out my wild ideas on students, faculty, and public audiences—with shout-outs to George Vrtis and Eunice Blavascunas and their colleagues for deep-dive visits at Carleton College and Whitman College. Thanks to Lawrence Culver and the American Society for Environmental History for their keynote invite, and to the Rachel Carson Center for publishing that early stab. Thanks always to Bill Cronon, whose mentorship endures in my work.

How can I possibly thank my readers enough, for their indispensable critiques, suggestions, and head-scratching: Cathy Gudis, Leila Philip, Allana Ross, Thea Mercouffer, Will Wright, Grant Price, Jonathan Price, Madelon Price, Phil Deloria, Sean Hecht, Giovanna Di Chiro, Lisa Eckstein, Nicole Seymour, Frances Wu, Veronique de Turenne, and Michael Alexander. Jane Paul, Stephanie LeMenager, Amy Lethbridge, and the social media hive mind slipped me crucial 39 Ways suggestions. My millennial-reader conscripts Micha Price, Jake Price, and Steven Ring did careful, brilliant readings—and tried, with Jessie Wu, Ian Price, Gabe Price, and Yishai Schuchalter, to save me from my "OK, boomer" mistakes. Many thanks, too, to Maxine Lipeles and Joel Goldstein for my home office away from home, and to Tiffanie Tran for her sound design sense, and to my indispensable on-call artist David Price (and to Charlotte, for being Charlotte).

Thanks always to Cindy Ott, a great friend and my first line of defense as a writer, who remains patient while I whinge and rant and rave and ultimately realize that she's right as always.

Many, many thanks to my agent Bonnie Nadell, who waited a great while. Many and deep thanks to my Norton editor Alane Mason: I'm so grateful you also waited, as you are a true editor's editor. Thanks to Mo Crist and Alexa Pugh for astute feedback, and to the entire Norton team, in whose hands this tiny odd duck of a thing has been safe and warm and much appreciated.

Warm thanks to Marty Gordon, and to all my family, for their love and support (yes, I'm finished), and a special shout-out to those who housed and fed me and my work in progress—Susan Green and Bobby Ring, Naomi Schacter and Jonathan Price, and Frances Wu and Grant Price. I've dedicated this book to my brothers' children and grandchildren, and I've written this book also for my godson Jonah and not-godson Asher, my cousin-niece Marissa and cousin-nephew Steven, my treasure trove of cousins' children, and my friends' children—all the stout-hearted young people in my life who keep me going.

Always thanks to Elmer Price, my Dad, whose faith, wit, love, intelligence, and integrity are lasting. Here's to the memory of my sweet brother David. And to Madelon Price, my legendary mad-scientist Jewish mother: I couldn't have done this without your good cheer and huge smarts, your nonstop cooking, and your unwavering support and love.